AF528589

Veterinary Histology

Veterinary Histology

Roy Mathew | Usurupati Samiyalu | T Anup Kumar

2026

Agri Horti Press

New Delhi-110 084

© 2026 Reserved

ISBN: 978-93-49690-25-7 (Hardbound)

Publisher's note: All Rights reserved under International Copyright Conventions. No part of this publication may be reproduced, stored in a retrieval system, or transmitted in any form or by any means, electronic, mechanical, photocopying, recording or otherwise without the prior written consent of the publisher and the copyright owner.

Published By

Agri Horti Press

New Delhi

E-mail: agrihortipress@gmail.com

Digitally Printed at : Replika Press Pvt. Ltd.

Preface

Veterinary histology plays a crucial role in understanding the microscopic structure and function of tissues and organs in animals. As veterinary medicine continues to evolve, the need for in-depth knowledge of tissue organization, cellular interactions, and pathological changes becomes increasingly important for diagnosing diseases, advancing therapeutic strategies, and improving animal health outcomes. This book, Veterinary Histology, is designed to provide an essential guide to the microscopic anatomy of animals, serving as a comprehensive resource for veterinary students, professionals, and researchers.

In the opening chapters, the book provides a foundational understanding of histology, starting with the basic concepts of cell biology and tissue organization. It explores the different types of tissues—epithelial, connective, muscular, and nervous—and their specialized structures and functions. Through detailed explanations and high-quality images, the book illustrates how these tissues contribute to the overall physiology of animals, helping readers to appreciate the importance of histology in the broader context of veterinary science.

The subsequent chapters delve deeper into the microscopic anatomy of various organ systems, including the integumentary system, digestive system, respiratory system, cardiovascular system, lymphatic system, nervous system, and reproductive system. Each chapter provides a detailed analysis of the tissue architecture and cellular components specific to these systems, with a focus on their roles in maintaining the health and function of the animal body. The book highlights how the histological structure of each organ system is closely tied to its physiological function and how deviations from normal tissue structure can indicate the presence of disease.

One of the key features of Veterinary Histology is its emphasis on the practical application of histological knowledge in veterinary practice. The book addresses how histological techniques, such as tissue preparation, staining methods, and microscopy, are used to examine tissue samples, diagnose diseases, and guide treatment decisions. It also explores the role of histology in understanding the pathogenesis of various conditions, from infectious diseases and neoplasms to metabolic and degenerative disorders. Case studies and examples from veterinary pathology are included to illustrate how histology is applied in real-world clinical settings.

Furthermore, the book examines the significance of histological research in advancing veterinary medicine. It highlights how cutting-edge techniques, such as immunohistochemistry, molecular histology, and digital pathology, are transforming the way histologists and veterinarians study animal tissues and diagnose diseases. These advancements open new avenues for research and have the potential to improve diagnostic accuracy, treatment options, and the overall understanding of animal health.

Veterinary Histology is not just a reference for students and professionals; it is a comprehensive resource for anyone involved in veterinary research, teaching, and practice. By bridging the gap between cellular biology and clinical veterinary medicine, the book provides a thorough understanding of the intricate relationship between the microstructure of tissues and the health of animals. It is an invaluable tool for enhancing diagnostic skills, improving treatment outcomes, and advancing knowledge in the field of veterinary histology.

Roy Mathew | Usurupati Samiyalu | T Anup Kumar

Contents

Introduction

The roles of veterinarians in society range from clinician to researcher, food inspector to enlisted military officer. Critical to the educational training of all of these veterinarians is the fundamental understanding of basic anatomy, both gross and microscopic. The knowledge of normal anatomy provides the basis of both physiology as well as pathology. Relating gross (e.g. physical exam findings) or microscopic changes (e.g. surgical biopsy) to specific disease processes first requires firm knowledge of "normal"; hence, disease is detected as an alteration of normal!

Teaching histology in the veterinary curriculum is the challenging process of conveying to the future veterinarian the clinical relevance of microscopic anatomy, which may not be inherently appreciable to the first-year veterinary student. The authors are fortunate to instruct professional students both in the preclinical curriculum and the senior clinical rotations where the relevance is readily apparent. When a senior student looks at a cytology of an osteosarcoma, they intuitively know that the neoplastic cells observed are osteoblasts with a small proportion of non-neoplastic osteoclasts. The same goes for ultrasonography of equine ovaries where they can extrapolate the "black hole" in an ovary as a large pre-ovulatory follicle containing an oocyte and lined by granulosa cells. From the instructor's perspective, it is easy to go overboard in terms of the amount of information conveyed to the student in this type of course, particularly if the instructor is passionate about a particular organ system! The modern veterinary medical curriculum is constantly changing and expanding. It is the instructor's responsibility to focus on not only conveying the necessary information, but packaging this information in the context of clinical medicine. For this reason, this text focuses on conveying the basic material required for understanding the microscopic anatomy to better understand the disease processes that they will learn later in the curriculum. Included is a basic overview of veterinary histology of commonly reviewed organ systems, with the goal of illustrating important concepts of cells, tissues, and organs in a manner that we hope is not only accessible to first-year veterinary students, but serves as a reference for clinical medicine and pathology..

Chapter-1

The Cell

The cell is the smallest and most basic unit of life and makes up all living organisms, whether unicellular or multicellular. This chapter will cover the cell.

Chapter Learning Objectives

By the end of this chapter, you should be able to do the following:

- ❖ Describe the structure and function of the phospholipid bilayer.
- ❖ Describe the important components of the nucleus.
- ❖ Describe the structure and function of smooth and rough endoplasmic reticulum.
- ❖ Understand the process in which the Golgi modifies and packages proteins.
- ❖ Describe the differences between lysosomes and perioxisomes.
- ❖ Understand the structure of mitochondria and how it relates to energy (ATP) production.
- ❖ Describe the differences between microfilaments, microtubules and intermediate filaments.
- ❖ Outline the processes of mitosis and meiosis.

Review Questions

By the end of this chapter, you should be able to answer the following:

- ❖ Describe the structure of the biological membrane, how is this structure related to
- ❖ the membrane functions?
- ❖ What is the difference between euchromatin and heterochromatin, how this difference is
- ❖ related to their functions?
- ❖ What are the structural and functional differences among microfilaments, microtubules, and
- ❖ intermediate filaments?
- ❖ What are the structural and functional differences among smooth ER, rough ER, and Golgi apparatus?

- ❖ How is lysosome associated with various forms of disease?
- ❖ Ryan Jennings and Christopher Premanandan
- ❖ How is the structure of mitochondria related to its function?
- ❖ Describe the structural changes that occur during mitosis and how these are related to the function?
- ❖ How is the genetic diversity created during meiosis?

Overview

The cell is the smallest and most basic unit of life and makes up all living organisms, whether unicellular or multicellular. In multicellular organisms, several cell types can interact to form specialized tissues and organs. Cells contain multiple organelles, or subcellular structures, that each have distinct structures and specific functions. Functions of these organelles are often dictated by the structure and are dependent on the location (tissue) of the cell and the physiological and disease status of the organism.

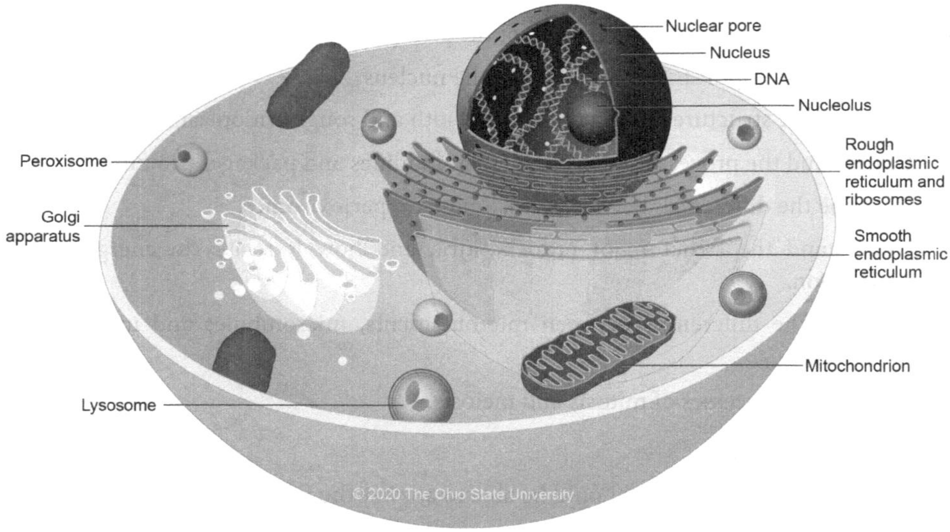

Cell membrane

The cell membrane acts as the barrier between inside and outside of the cell. It is composed of an asymmetric phospholipid bilayer with embedded proteins and carbohydrates and is approximately 7 nm thick. The phospholipids in the membrane can move through lateral diffusion, flexion, or rotation, a process called the fluid mosaic model. The asymmetric bilayer serves an important function for the cell by regulating molecular diffusion across the membrane based on charge and size. The two general categories of proteins present in the cell membrane are referred to as integral or peripheral proteins. Integral proteins form hydrophobic bonds with lipids and other integral proteins. These proteins penetrate partially or through lipid bilayer and may be glycoproteins (oligosaccharides). Peripheral proteins are less hydrophobic than integral proteins. They may associate with lipid polar headgroups or integral proteins via H-bonds or ionic interactions.

Some proteins exist as receptors located within the membrane, with functions ranging from activating downstream intracellular signaling when bound by ligand or acting as channels to allow ions to flow into or out of the cell. Other functions of the membrane are to subdivide the cytoplasm within the cell and increase surface area of the cell. Within the membrane are structures named lipid rafts. These areas are enriched in cholesterol, sphingolipid, and proteins. Lipid rafts are important to the cell for signal transduction across the membrane.

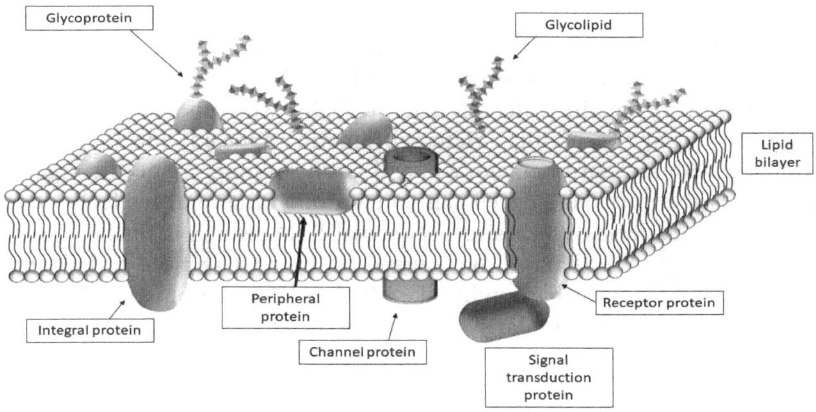

Diagram of cell membrane

Nucleus and nuclear structures

The nucleus contains and confines the genetic material of a eukaryotic cell. Prokaryotic cells do not have a defined nucleus, but rather have genetic material within the

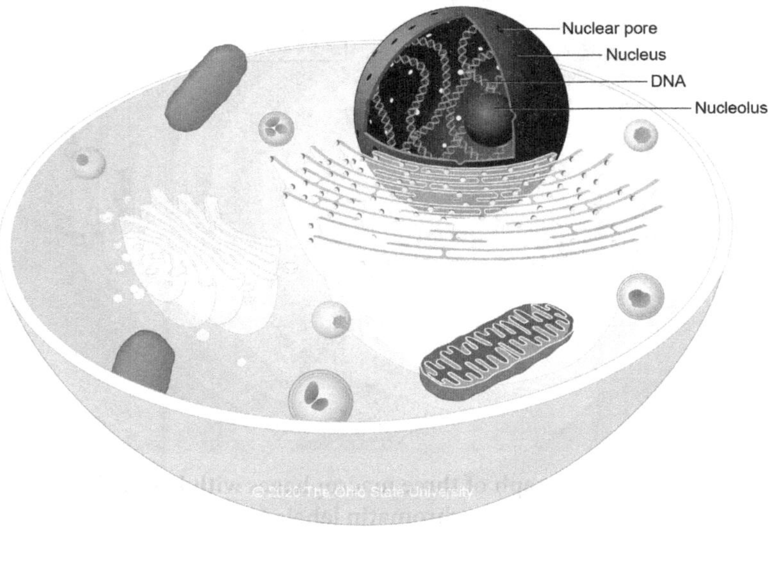

cytoplasm. The nuclear envelope is the cellular component that surrounds and defines the nucleus. This structure is composed of two membranes that are continuous with the rough endoplasmic reticulum that will be further discussed later in this chapter. Within the nuclear envelope are nuclear pores, donut-shaped symmetric ring structures that allows selective transport of materials such as RNA or proteins, lipids, and carbohydrates into or out of the nucleus. However, a nuclear localization sequence is necessary before transport can occur. The fibrous lamina is the third component of the nuclear envelope. The fibrous lamina is composed of lamin and membrane-associated proteins and is found on the inner surface of the inner nuclear envelope. This component is responsible for overall nuclear stability and organizing nuclear events such as DNA replication and mitosis.

Within the nucleus, DNA and histones are packaged into either euchromatin or heterochromatin. Euchromatin is loosely packed DNA, enabling genetic transcription to occur. Heterochromatin is densely packed DNA, which is inactivated and prevents transcription from occurring. Heterochromatin appears more basophilic than euchromatin with standard hematoxylin and eosin staining. A specific organelle within the nucleus is the nucleolus. A cell can contain one or several nucleoli, depending on its activity. The nucleolus synthesizes ribosomal RNA, which is further assembled into ribosomes, an organelle that will be covered in the next section. In light microscope, nucleolus appears intensely basophilic, like a blob of heterochromatin. However, the bulk of nucleolus is RNA so the similarity is very superficial.

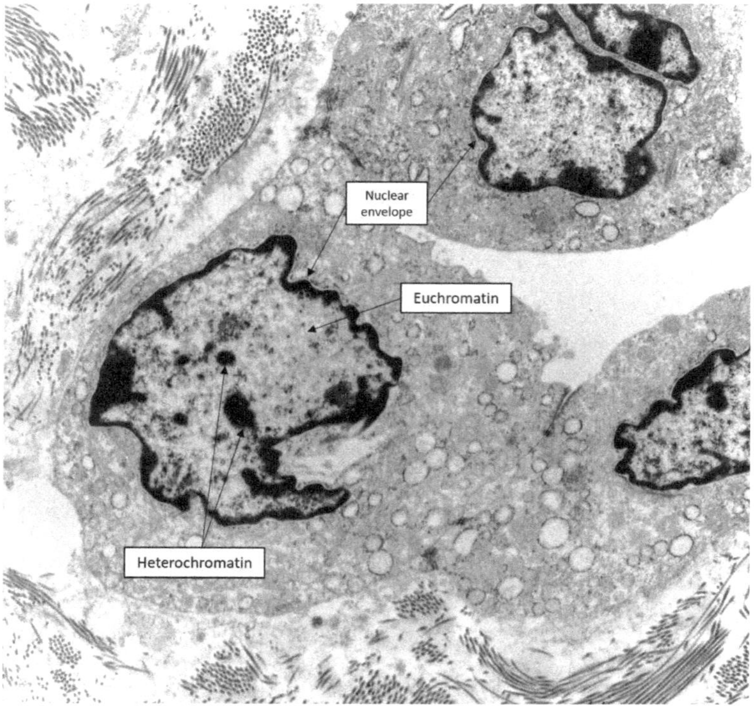

Electron photomicrograph of three macrophages with heterochromatin and euchromatin labeled.

Endoplasmic reticulum

Rough endoplasmic reticulum (rER) is composed of a membrane (ER) that is a continuation of the nuclear envelope and ribosomes. RER is arranged in a sheet-like pattern and functions to synthesize membrane and secretory proteins. Ribosomes can also be free within the cell to synthesize cytoplasmic proteins. They are composed of protein subunit structures that aid in the conversion of an mRNA sequence into an amino acid sequence, thereby producing a polypeptide chain that will fold into a three-dimensional protein. Ribosomes are composed of two asymmetric pieces: large (60S) and small (40S) subunits. The large (60S) subunit is made of three pieces of RNA (28S, 5.8S and 5S) comprising 65% of subunit, and many other proteins, comprising 35%. The small (40S) subunit made of one piece of RNA (18S) and many other proteins. The specificity of protein-RNA interactions is so high that a test tube filled with the right RNA and proteins will self-assemble into a ribosome. Ribosome, mRNA, tRNA, and enzymes may associate free in cytoplasm and make proteins. Cells use this format when synthesizing cytoplasmic proteins.

In contrast, smooth endoplasmic reticulum (sER) is both structurally and functionally different than rER. Firstly, sER does not contain ribosomes and, therefore, is not involved in protein synthesis. Within the cell, sER appear as tubules that can either be parallel, similar to rER, or as tangles or a "lace- like" pattern. The sER has several important functions throughout the body. In the endocrine system, this organelle is involved in the biosynthesis of steroids, including testosterone and estrogens, from cholesterol. In the liver, it is highly involved in detoxification of drugs from the body. A third function of the sER is its relationship to the contraction of smooth and striated muscle. In muscle, the sER has the specific name of sarcoplasmic reticulum, which stores intracellular calcium ions. Release and sequestration of calcium ions has a direct effect on muscle contraction and relaxation, which will be covered in greater detail in in the muscle chapter.

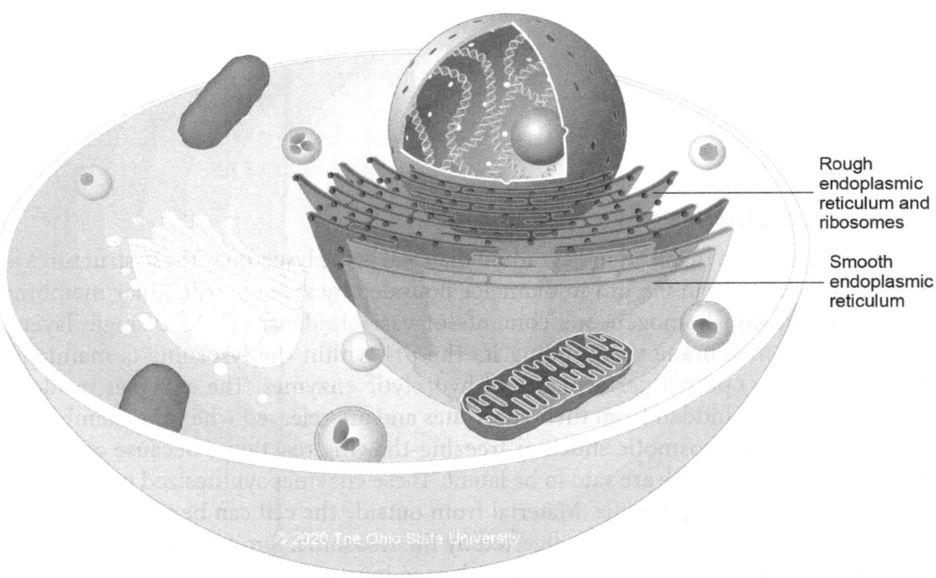

Golgi Apparatus

Once proteins are synthesized by ribosomes and the rER, they are transported to the Golgi apparatus where further processing takes place. This represents the end of the protein synthesis chain. The Golgi apparatus performs two main functions in relation to protein processing and packaging. First, it attaches sugars to proteins to form glycoproteins in a process called glycosylation. It receives the newly formed polypeptides from the rough ER. Due to resident glycosyltransferase enzymes, the Golgi can attach sugars to the proteins making glycoproteins. Second, the Golgi then condenses the proteins, packaging them into membrane bound secretion granules.

The Golgi apparatus has a distinct cup-shaped structure with cisternae, or flattened membranes, arranged in parallel sheets. Proteins enter the Golgi apparatus on the convex, or forming face, side and exit on the concave side or maturing face.

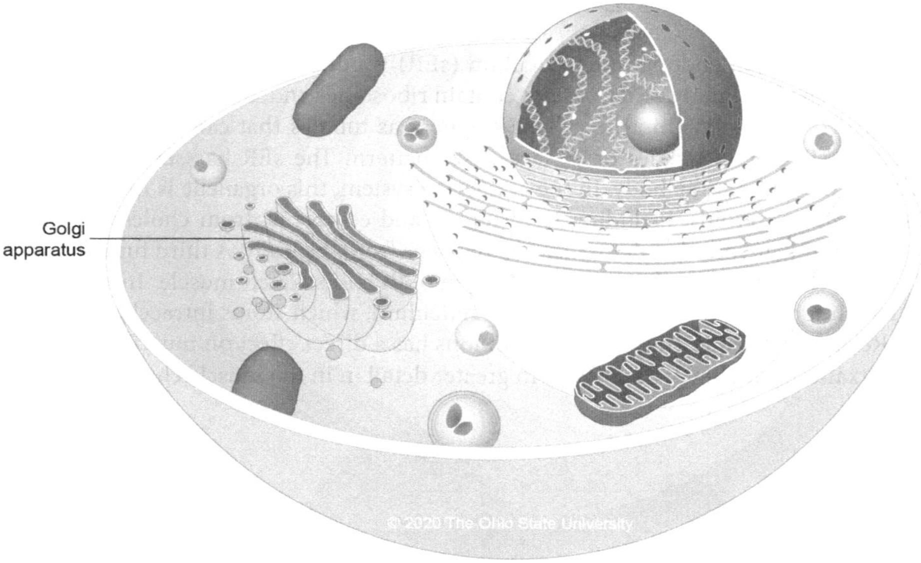

Radation Organelles

One of the organelles involved in degrading material is the lysosome. These structures are spherical particles about 0.5 μm in diameter bounded by a single trilaminar membrane containing (usually) homogeneous contents of variable density with a single layer of semipermeable membrane surrounding it. The pH within the lysosome is maintained between 4.5-5 for optimal performance of hydrolytic enzymes. The enzymes inside the intact lysosome are hidden from their substrates and are released when the membrane is broken by detergents, osmotic shock or freezing-thawing reactions. Because of this, the enzymes inside a lysosome are said to be latent. These enzymes synthesized in the ER and packaged in the Golgi apparatus. Material from outside the cell can be ingested through phagocytosis or pinocytosis and is digested by the lysosome. Similarly, the lysosome can digest material from inside the cell that is either no longer needed or degraded through a process called autophagy. This process allows for orderly digestion of cellular components without incurring damage to the entire cell. The digested components can then be recycled.

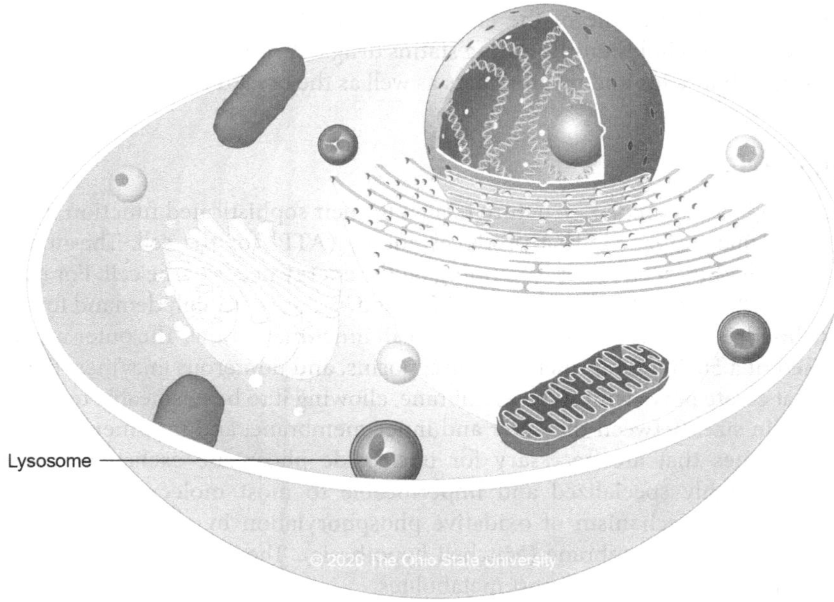

A second organelle involved in degradation is a peroxisome or microbody. These are 0.5 μm spherical membrane-bounded organelles that closely resemble a lysosome. Smaller "microperoxisomes" are found in some types of cells, e.g., intestine. These structures are

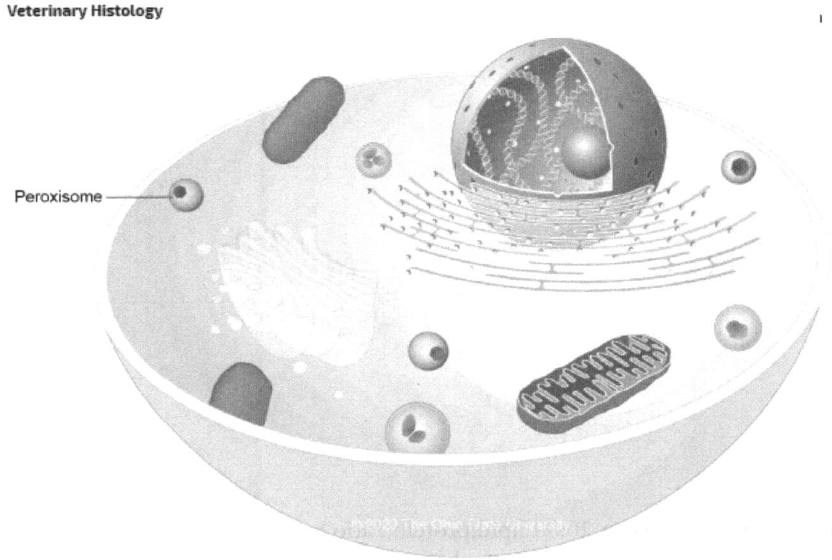

small and spherical with a single semipermeable membrane. Unlike lysosomes though, peroxisomes have a crystalline core. Peroxisomes contain catalases and oxidases, which use oxidation to break down H_2O_2 and excess fatty acids. They also participate in the cholesterol biosynthesis. One of the enzymes involved, HMG- CoA reductase, is the target of the popular cholesterol-lowering statins drugs. Perioxisomes also are involved in the synthesis of bile acids and myelin lipids as well as the breakdown of excess purines to uric acid.

Mitochondria

Mitochondria have a complex structure related to their sophisticated function in the cell and is the primary organelle that produces energy (ATP) for the cell. The number of mitochondria within a cell varies depending on the energy needs of the cell. For example, a muscle cell will have a high number of mitochondria due to its high demand for energy. The mitochondria is composed of an outer and an inner membrane. The outer membrane is composed of a 50:50 lipid:protein mixture, porins, and numerous enzymes. Porins are proteins that create pores within the membrane, allowing it to be permeable to molecules up to a certain size. Between the outer and inner membrane, and intermembrane space contains enzymes that are necessary for nucleoside phosphate exchange. The inner membrane is highly specialized and impermeable to most molecules that, in turn, is involved in the mechanism of oxidative phosphorylation by allowing ionic charge separation across this membrane (Mitchell hypothesis). Therefore, specific carriers are needed for the entry and exit of most metabolites.

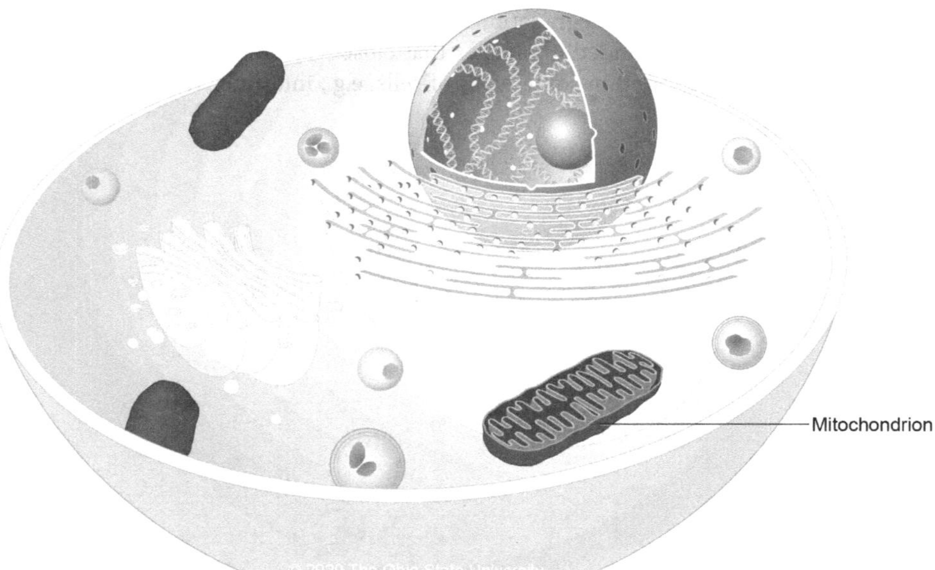

The inner membrane has a 20:80 lipid:protein ratio and contains the enzymes involved in the respiratory electron transport chain and phosphorylation of ADP. Therefore, the inner membrane is the primary site of ATP synthesis. Cristae are projections of the inner membrane into the mitochondrial matrix; the number of cristae present is directly related to the amount of ATP synthesized. The enzyme ATPase is attached to the matrix surface of cristae and directly couples the electron transport chain to ADP phosphorylation. Other molecules attached to the cristae include cytochrome C on the outer surface, which has a role in the electron transport chain and apoptosis, and succinic dehydrogenase on the matrix surface. Within the mitochondrial matrix are enzymes involved in metabolism of nucleic acids, lipid synthesis, protein synthesis, and the Kreb's Cycle. Mitochondria also play an important role in the regulation of programmed cell death (apoptosis).

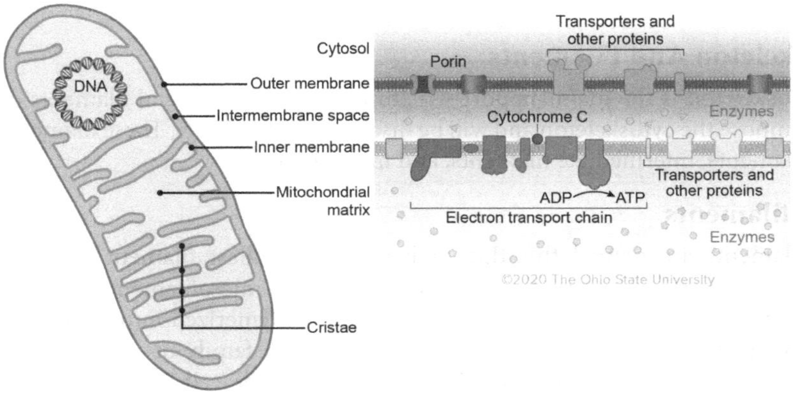

Mitochondria contain genetic material that is inherited from the mother and is found in the mitochondrial matrix as a circular double stranded molecule. However, only

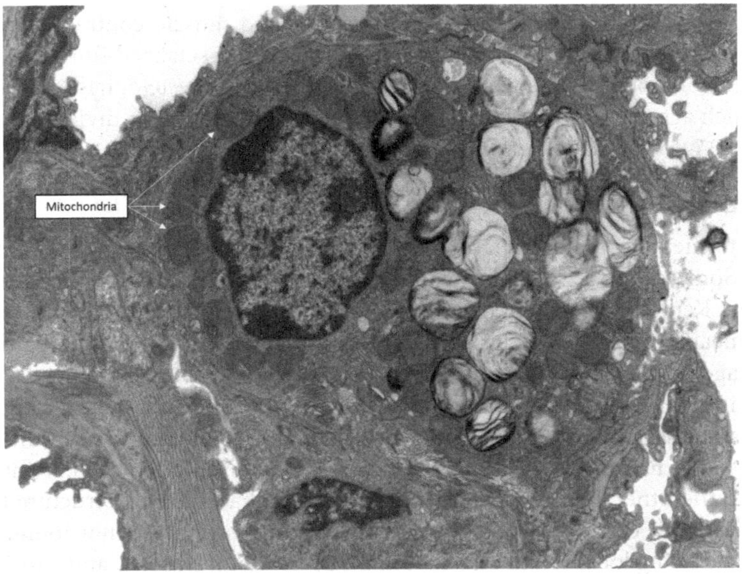

Electron photomicrograph of a Type II pneumocyte from the pulmonary parenchyma. Mitochondria can be observed at the periphery of the cell predominantly around the nucleus. Cristae are visible as slightly electron dense lines spanning the mitochondria.

mitochondrial RNAs and a small number of inner membrane proteins are translated and transcribed from this DNA. Most mitochondrial proteins are translated and transcribed from nuclear genes. Therefore, they are synthesized in the cytoplasm of the cell and imported through membrane protein translocase complexes across the outer and/or inner membranes into the mitochondria. The presence of DNA in mitochondria provides evidence for how mitochondria may have developed in the cell, a process called mitochondrial biogenesis. It is theorized that mitochondria are derived from aerobic endosymbiotic bacteria, meaning bacteria that entered and remained to provide a benefit to the host cell. Another piece of evidence to support this theory is that mitochondria divide through a process called binary fission, similar to bacteria.

Cytoskeleton And Filaments

The cytoskeleton is the main support structure for the cell. Several filaments of various sizes compose the cytoskeleton and will be discussed individually. These filaments include microfilaments, intermediate filaments, and microtubules.

Microfilaments

Microfilaments are named this due to being the smallest filament structure of the cytoskeleton, usually around 6 nm in diameter. One microfilament is actin filament or F-actin, which is composed of actin monomers polymerized into a double-stranded helix. While not a microfilament per se, myosins are a family of proteins that interact with actin to perform various motor activities within the cell. Myosins contain three domains. The head domain binds actin and hydrolyzes ATP to generate movement, the neck domain can bend to accommodate the movement of the molecules, and the tail domain can bind other molecules for transport. Actin and myosin are highly involved in cell motility, organelle and vesicular transport, and muscle contraction as they are the main components of the sarcomere in muscles. A specialized function of an actin and myosin compound is the contractile ring formed during telophase of cell division, during which actin and myosin line up in the middle of the cell to divide the cytoplasm (cytokinesis) into two daughter cells.

Intermediate Filaments

Several intermediate filaments are found within cells, with sizes around 10 nm in diameter. Some examples of intermediate filaments include keratin, vimentin, desmin, and lamin. Keratin is formed from monomers that combine into bundles, giving it an impermeable quality. Due to its toughness, it functions to protect epithelial cells from damage, with one of its main locations in the body being the outer layer of skin. Vimentin is also formed from monomers that interact to form a helical structure. This filament attaches to several organelles in the cell to anchor them into position. It also has a highly flexible nature, which suggests that it is responsible for cellular integrity and shape. Like the other intermediate filaments, desmin has a helical structure that allows it to interact with other molecules. However, desmin is typically not found in all cell types, instead being localized to cells that make up smooth, skeletal, and cardiac muscle. Therefore, it plays a role in muscle contraction. As discussed previously, lamin is found on the inner surface of the inner nuclear envelope and makes up the fibrous lamina in the nucleus. It is also a helical filament that is involved in nuclear pore formation and ensuring proper disassembling and reassembling of the nuclear envelope during and after mitosis. Lamins are also a target during apoptosis.

Microtubules

The final type of filament that makes up the cytoskeleton is the microtubule. Around 25 nm in diameter, it is composed of an α and β heterodimer of tubulin that polymerize end-to-end to form protofilaments. Each tubulin dimer normally binds one nonexchangeable molecule of GDP and one exchangeable molecule of GTP. In addition, each dimer has binding sites for the drugs colchicine (used in the treatment of gout) and vinblastine (used in cancer chemotherapy). The microtubule makes up different structures depending on the number and organization of protofilaments. For example, central pair and 9 outer doublet microtubules (9+2 arrangement) are the main structural units (axoneme) of flagella and cilia. They also interact with the ATPase dynein and ATP resulting in transduction of chemical energy and the sliding of outer doublets relative to each other. This results in the bending movements of the organelles. In many asymmetrical cells, microtubules are positioned in a way that stiffens or supports the cytoplasm, e.g., long spike-like pseudopods of some protozoa; axons and dendrites of nerve cells; elongating embryonic cells. Microtubules are also involved in the mitotic apparatus during cell division and the directed transport of organelles and molecules along a nerve cell axon (axoplasmic flow).

The centriole is a nonmembrane organelle which exists in pairs in each cell. Each centriole measures 0.15 x 3-5 μm and is a hollow cylinder closed at one end. The walls of the

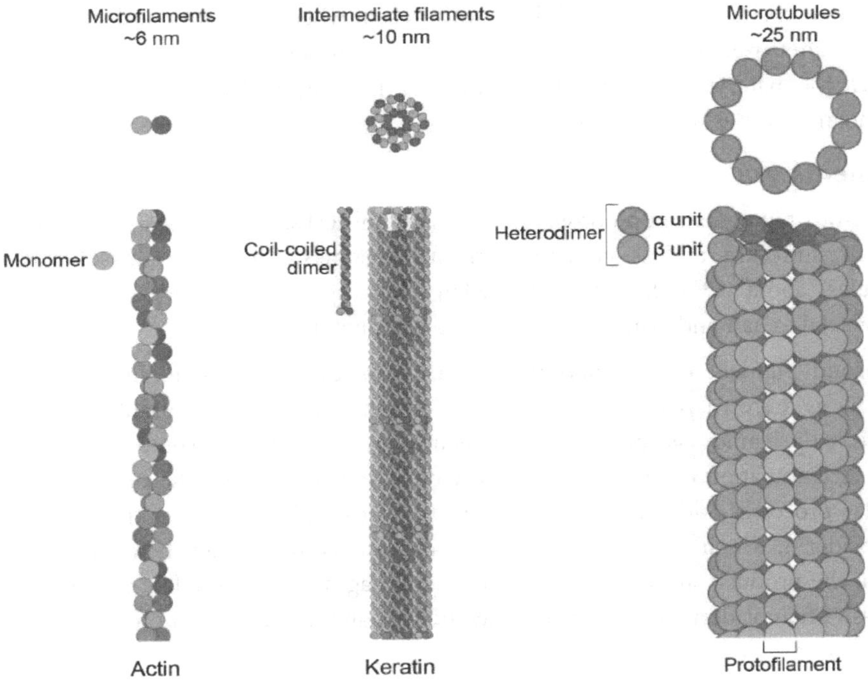

cylinder are composed of nine sets of triplet microtubules, each tubule similar in dimension to a microtubule (250 Å in diameter). The pair of centrioles usually lie near the Golgi apparatus in resting cells, but they divide and play a role in establishing cell polarity and the spindle during cell division (but no direct contact of spindle fibers with centrioles).

Cell inclusions

Cell inclusions are considered various nutrients or pigments that can be found within the cell, but do not have activity like other organelles. Examples of cell inclusions are glycogen, lipids, and pigments such as melanin, lipofuscin, and hemosiderin.

Glycogen

Glycogen is the long-term storage unit of glucose within the cell, typically in liver and muscles. Glucose molecules are connected by α(1-4) linkages and branched off by α(1-6) linkages to form the complex structure of glycogen, which also aids in rapid breakdown. Glycogen can be visualized in tissue using a periodic acid-Schiff (PAS) stain. Under an electron microscopic, two forms of glycogen exist. The first is a single spherical 15-30 nm particle (β–particle) that stain densely with lead. The second are aggregates of small particles (β-particles) called rosettes (α-particle).

Lipids

Lipid spheres in tissues are caused by an accumulation of triglycerides and appear as perfectly spherical structures. Due to processing, lipids cannot be visualized in paraffin-embedded tissues; it is assumed lipid was present when clear, spherical structures are present. On frozen tissue, lipids can be visualized using stains such as Sudan Black or Oil Red O. The material is prominent in adipocytes (fat cells), and may be found in any cell. The quantity varies with metabolic state.

Pigments

Numerous pigments can be observed in tissues and cells. Melanin is a brown pigment in hair and skin and is contained in melanosomes. These are dense ellipsoidal granules (about 0.3 x 0.7 μm) produced by specialized cells called melanocytes and are transferred to hair and skin cells to produce a pigmentary change.

A second pigment often found in cells and tissues is lipofuscin. Lipofuscin often appears as a brown-yellow color that can be autofluorescent and accumulates over time, giving it the name "age pigment". It is mainly found in the lysosomes of postmitotic cells. Accumulation of lipofuscin can have detrimental effects on the cell by disturbing cellular processes and leading to a degradation of cellular activity. A third pigment is hemosiderin, which is a brown pigment that contains iron. When hemoglobin breaks down, the products are phagocytized by macrophages and accumulate within the cell. Excessive hemosiderin accumulation may indicate an increase in hemolysis of red blood cells.

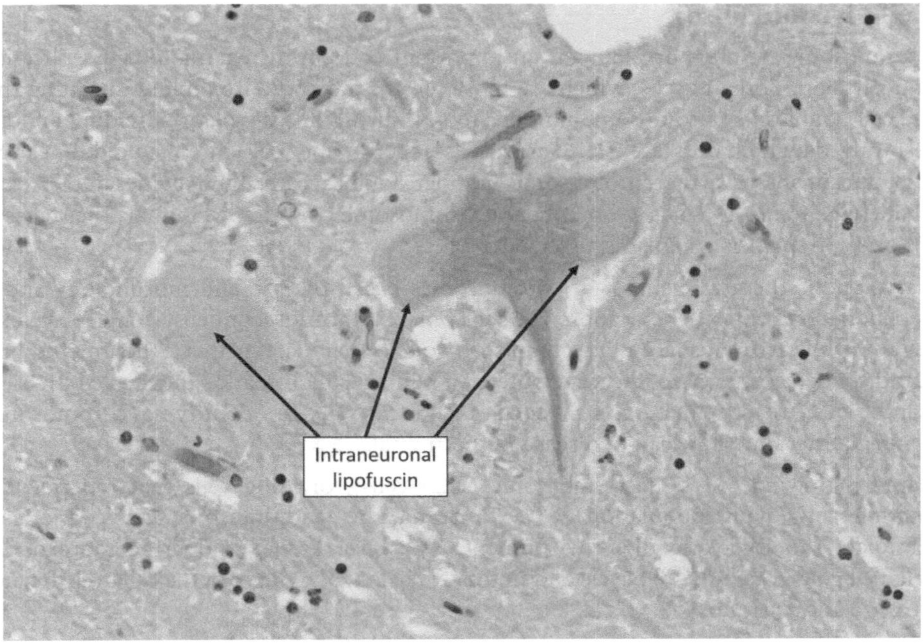

400X photomicrograph of two neurons from the cerebrum of an older canine. The faint yellowish brown pigment is lipofuscin.

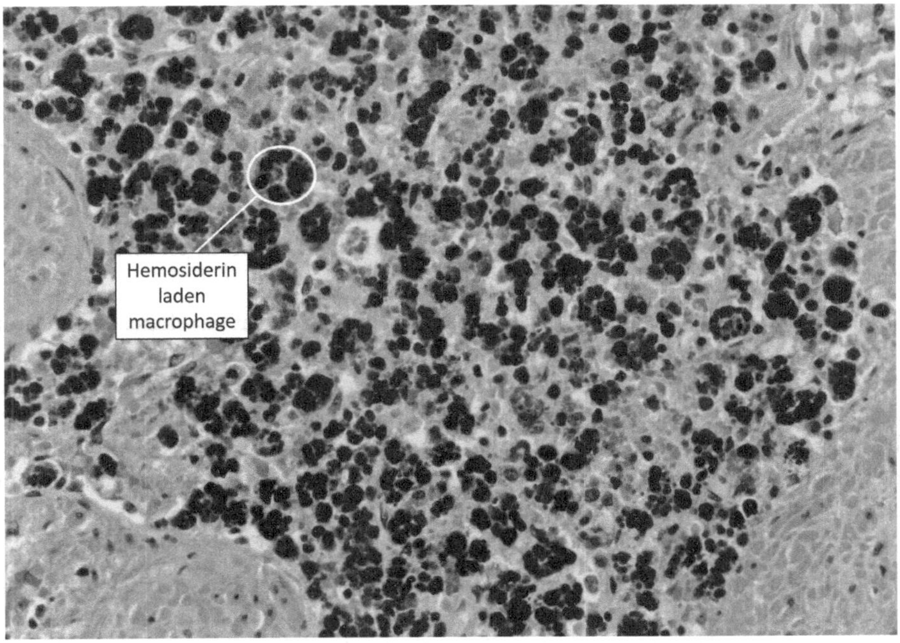

400X photomicrograph from a canine spleen. This area contains an aggregate of macrophages that contain large amounts of intracytoplasmic dark brown pigment (hemosiderin).

Cell division

Chromosome number, or ploidy, is an important concept in regards to cell replication and division. Somatic cells, which are most cells in the body, are diploid, meaning that the cell doubles its chromosome number to 4N during mitosis before dividing and the resulting daughter cells are 2N. Germ cells (sperm and egg) are haploid, meaning that during meiosis, the chromosome number doubles to 4N, then divides into 2N, and divides again into 1N. N is the number of chromosome types in the animal. The stages of mitosis and meiosis will be discussed later.

Karyotype is the arrangement of chromosomes based on size and morphology and is dependent on centromere location. The centromere is the most condensed and constricted region of the chromosome that connects two chromatids together. The microtubule spindle fibers connect to the centromere to pull apart chromatids during cell division. Karyotypes are specific for each species and can aid in identifying chromosomal abnormalities that may lead to disease in the animal.

Rapidly growing cells (e.G., Embryonic, stem cells), dividing cells that repopulate a continuously dying population (e.G., Bone marrow and intestinal cells), and cells stimulated to divide by the environment (e.G., Liver, kidney, endocrine and exocrine cells) all go through the cell cycle and mitotic division. The stages of the cell cycle include g1, s, g2 and m (mitosis).

The cell cycle

Cells in g1 (g is for "gap") are rapidly synthesizing rna and protein, enlarging their nuclei and cytoplasm from the last cell division. Dna replication (synthesis) occurs in the s period. The cell will double its dna in an exact copy in about 7 hours. Interestingly, not all dna is replicated at once, but in a cell-specific sequence (to preserve ploidy, chromosome number goes from 2n to 4n, then will be reduced to 2n again at m). A second gap (g2) takes place that is another checkpoint prior to m. Cells with polytene chromosomes and polyploid cells (e.G. Mammalian liver cells) remain in this period indefinitely. G2 is about an hour in rapidly renewing tissues.

Mitosis, or the process of cellular division, can be divided into four distinct phases, each with specific changing cellular structures.

Prophase is the first phase and involves breakdown of the nuclear envelope, disappearance of the nucleus, and condensation of chromatin. Chromatin coils, supercoils, and makes loops to become a chromosome. Separate copies of genome become visually segregated into two sister chromatids, remaining attached to each other by a protein called cohesin at the centromere (specific dna sequence). As discussed previously, the inner nuclear envelope is surrounded by lamin. The lamin disintegrates when it receives signaling from mitosis promoting factors. The envelope then is contained in numerous vesicles, which is retracted into mitotic er until two new cells are formed.

After disintegration of the nuclear envelope, the cell is in metaphase. During this phase the completely condensed chromosomes (each pair of sister chromatids) align themselves in a plane normal to a line joining the two pairs of centrioles. This plane is called metaphase plate. Some microtubules of spindle are inserted to the kinetochore (the disk-like protein attached to the centromere of chromosomes), others (continuous fibers) stretch from pole to pole, avoiding the chromosomes. Still a third class of microtubules, which originate at the centrioles, radiate away from the metaphase plate and the chromosomes, forming the aster. Note that the homologous chromosomes (chromosome pairs that each inherited from maternal and paternal sources) act independently in mitosis. Each is doubled into two sister chromatids, each will divide and be represented in the daughter cells.

Eventually, each chromatid is separated from its sister during anaphase. The centromeres holding the sister chromatids split in half, and each chromatid, now called a daughter chromosome moves in opposite directions to the asters. The splitting to the entire genetic message into two equal halves has been accomplished (karyokinesis: the dividing of the nucleus).

The process of uncoiling all chromosomes and reforming nuclear envelopes comprises the final stage, telophase. At this time, the nucleoli reform and the cytoplasm divides (cytokinesis), aided by a sub-plasma membrane purse-string of actin and myosin filaments, the contractile ring (creates cleavage furrow). Once two distinct cells are formed, the nuclear envelope reassembles. Membrane from the ER is used to form the envelope, with scaffolding from lamin forming the inner fibrous layer. The chromosomes in the resulting cells tend to be identical to the cell from which they originated.

Meiosis

Conversely, meiosis is a process designed to result in genetic variation. This variation occurs through both homologous recombination and random segregation of the chromatin. Meiosis is the process in which germ cells (sperm and egg) are formed. Therefore, having genetically distinct cells compared to the parent cell ensures that there will be genetic diversity in offspring. Also, germ cells have N number of chromosomes due to the second cell division that occurs during meiosis. When the sperm fertilizes the egg, the resulting cells in the offspring will have 2N number of chromosomes.

During meiosis, the ploidy is reduced from diploid (2n) to haploid (1n) in production of eggs and sperm. With the successful union of egg and sperm, the normal diploid (2n) complement of chromosomes is restored. To achieve this reduction in ploidy, the chromosome number is first doubled, as in mitosis, but then divided twice, so that there are four haploid daughter gametes from one diploid parent cell. Meiotic prophase is complex. What is important is that the homologous chromosomes do not act independently, but pair together (synapsis) in complex patterns. The sites in pairing between two homologous chromosomes is called synaptonemal complexes. In addition, chiasma (specific points where two duplicated homologous – maternal and paternal – chromosomes are physically connected) can form in two chromatids, one from each of the homologous chromosomes, where actual exchange of chromosome segment takes place. These structures allow recombination and crossing over to occur, the exchange and mixing of genetic information between maternal and paternal genes.

Metaphase I and Anaphase I proceed as in mitosis except that the homologous chromosomes pair on the metaphase plate. Anaphase I separates the homologous chromosomes, each one composed of a pair of sister chromatids. Random segregation of homologous chromosomes to two daughter cells has a chance of creating 2N variations. So at the end of Anaphase I, the maternal and paternal chromosomes are now spatially separated, except for those genes mixed during crossover. There is a brief interphase here, but no new DNA is replicated. Each daughter cell now undergoes a second division, a Metaphase II and Anaphase II. At this time each bivalent splits into two sister chromatids, which separate as in mitosis. None of daughter cells have identical sets of chromosomes.

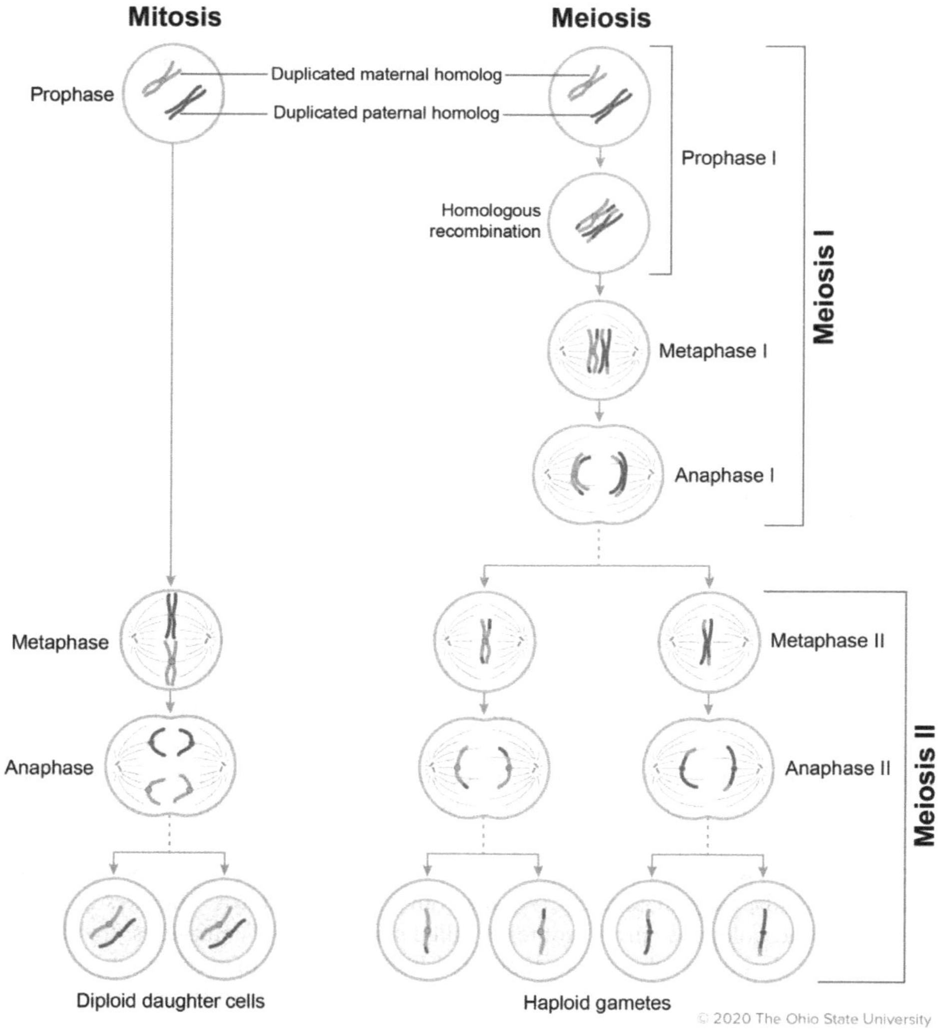

Chapter-2

Epithelium

All organs of the body are composed of combinations of the four basic tissues, epithelial, connective, muscular and nervous. This chapter will cover epithelium.

Chapter Learning Objectives

By the end of this chapter, you should be able to do the following:

- Describe the common features of epithelia.
- Classify the various types of epithelia using the correct terminology.
- Discuss the organization and functions of the intercellular junctions.
- Describe the types, organization and functions of glands.
- Based on the function of the various epithelial cells, predict the intracellular components and
- morphology of each epithelial cell type.
- Review Questions
- By the end of this chapter, you should be able to answer the following:
- What are the functions of epithelium?
- What are the common features between epithelial types?
- What does basolateral mean?
- How is epithelium classified in terms of cell layers?
- What are the differences between glandular and non-glandular epithelium?
- What are some modifications of epithelial cells and what are their functions?
- What is the difference between a desmosome and hemidesmosome?
- How is epithelium classified according to cell shape?
- What is cornification?

Overview

All epithelial tissue rests on a basement membrane. The basement membrane acts as a substrate on which epithelium can attach to as well as grow and regenerate after injuries. Epithelial tissues have a nerve supply, but they do not have a blood supply.

Therefore, epithelia must be nourished by substances diffusing from vascular structures in the underlying tissue.

The primary functions of epithelial tissues are:

1. To protect the tissues that lie beneath it from injury such as desiccation, invasion by pathogens, and physical trauma
2. The regulation and exchange of molecules between the underlying tissues another compartment
3. The secretion of hormones into the blood vascular system, and/or the secretion of sweat, mucus, enzymes, and other products that are delivered by ducts

Epithelium lines both the outside (skin) and the inside cavities and lumina of bodies. For instance, the uppermost layer of skin is composed of dead stratified squamous, keratinized epithelial cells while the outer lining of many organs is comprised of mesothelium.

The type of epithelium that is present in an organ system depends on the function of the system. Surfaces that require a water resistant surface and protection from abrasion often are composed of keratinized stratified squamous epithelium (skin) or nonkeratinized stratified squamous epithelium (oral cavity, esophagus,rectum). Other

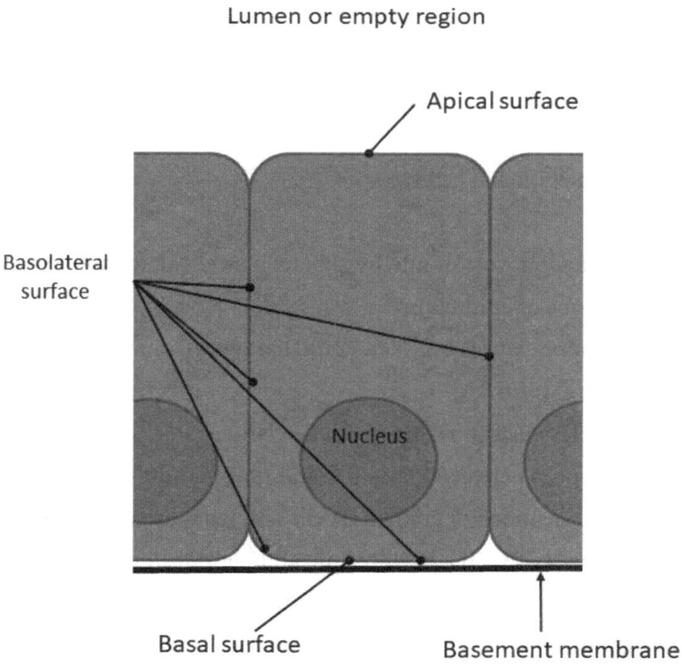

The apical surface is associated with a lumen or a free space. The basal surface is associated with the basement membrane. The lateral surface is the area in which an epithelial cell touches the adjacent one. The basal and lateral surfaces are often referred to as the basolateral surface.

surfaces that separate body cavities within the body are lined by simple squamous, columnar, or pseudostratified epithelial cells. Some types of epithelium can form solid masses of tissue or the majority of an organ. Hepatocytes comprise a majority of the hepatic parenchyma and are considered epithelial cells. Endothelium (the inner cell lining of vascular structures) is a specialized form of epithelium. Another type of specialized epithelium, mesothelium, forms the outside surface layer of the pericardium, pleura, and peritoneum. Some epithelial cells are organized into secretory structures at different locations in an animal's body known as glands.

All epithelial cells types share the following characteristics:

1. Close apposition of cells.
2. Free surface of epithelial cells is adjacent to the space.
3. Basal surface is adjacent to connective tissue.
4. Sheets of epithelial cells may be modified into tubes forming glands.
5. Absence of blood vessels within epithelial layer

Morphological classification of epithelia

Epithelium is further characterized by several physical features. These include :

1. **The number of layers of cells**: an epithelium with only one layer is referred to as *simple*. When there are more than one layer, the epithelium is referred to as *stratified*. A confusing exception to this is *pseudostratified epithelium which* appears to be more than one cell thick since the nuclei lie at different heights within the cell, but all cells are in contact with the basement membrane.
2. **Shape of cells at *free surface*:** Shapes of epithelial cells include *squamous* (flattened), *cuboidal* and *columnar*.
3. **Function of the epithelium**
4. **Surface modifications** (if present): Surface modifications include *cilia and microvilli*. Their presence often depend on the requirements of the tissue location where the epithelial cell resides.

Classification By Number Of Layers Of Cells

Simple epithelium

Simple epithelium is a single layer of cells with every cell in direct contact with the basement membrane. It can serve many functions, including absorption, filtration and protection. The thinness of the epithelial barrier facilitates easy movement of molecules across into another compartment.

In general, simple epithelial tissues are classified by the shape of their cells. The four major classes of simple epithelium are as follows:

1. **Simple squamous**: this type is found in lining areas where passive diffusion of gases occur, such as walls of capillaries and alveoli of the lungs. Endothelium is the special

term used for simple squamous epithelium that lines the inside surface of a vascular structure.

2. **Simple cuboidal**: these cells may have secretory, absorptive, or excretory functions. This type of epithelium can be seen in the tubules of kidney, and the secretory ducts of the pancreas and salivary gland.
3. **Simple columnar**: found in areas with extremely high secretory activity (as in the inside lining of the stomach), or absorptive activity (as in small intestine). This type of epithelium often possess apical modifications such as the microvilli in the small intestine, or cilia found in some areas of the female reproductive tract).
4. **Pseudostratified epithelium**: this type of epithelium is almost exclusively located in the larger respiratory airways of the nasal cavity, trachea and bronchi. For this reason, it is often also called respiratory epithelium.

There are two notable examples of simple squamous epithelium that have unique names. The first, endothelium, refers to the simple squamous epithelium that lines in the inside surface of all vascular structures. Endothelium is a critical component of vascular structures that will be discussed in the cardiovascular section. The second, mesothelium, refers to the simple squamous epithelium that lines the outside surface of many organs in the abdominal and thoracic cavities. It forms a component of the serosa, the lining of abdominal and thoracic organs as well as the parietal peritoneum and pleura.

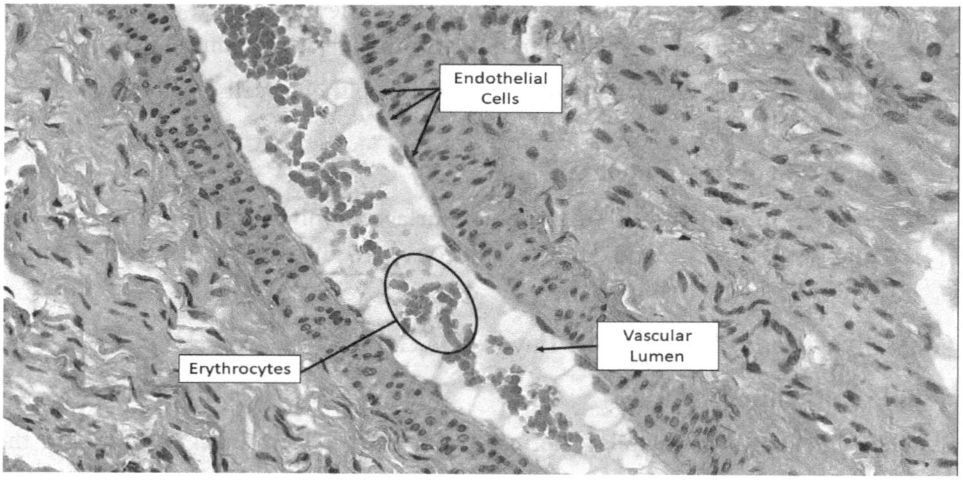

200X photomicrograph of an arteriole demonstrating endothelial cells

Stratified epithelium

Stratified epithelium differs from simple epithelium in that it is multilayered. It is therefore found where body linings have to withstand mechanical or chemical insult such that layers can be abraded and lost without exposing subepithelial layers.

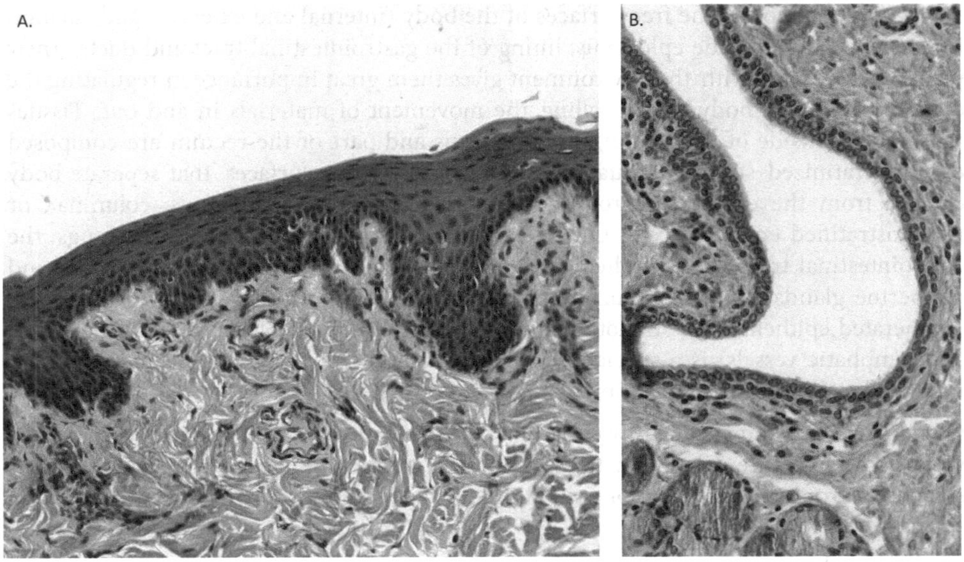

400X photomicrographs of stratified epithelium. A. Stratified epithelium from the inside of the lip (mucocutaneous junction). Stratified cuboidal epithelium from a duct. Stratified cuboidal epithelium is not frequently observed.

Classification by shape of the cells at The free surface

The three principal shapes associated with epithelial cells are—squamous, cuboidal and columnar.

- **Squamous epithelium** is composed of cells that are wider than their height (flat and wide).
- **Cuboidal epithelium** is composed of cells whose height and width are approximately the same (cube shaped).
- **Columnar epithelium** is composed of cells taller than they are wide.

By layer, epithelium is classed as either simple epithelium (one cell thick) or stratified epithelium (two or more cells thick), and both can be made up of any of the cell shapes. When epithelium is stratified, it is classified by the shape of the cells at the top or uppermost layer. A notable exception is pseudostratified epithelium. In this type of epithelium, taller simple columnar epithelial cells appear to have several nuclei at different heights, giving the appearance of stratification. However, in this type of epithelium, all the cells are attached to the basement membrane.

Urothelium (formerly called transitional epithelium) is another type of unique epithelium in which cells can change from squamous to cuboidal, depending on the amount of tension on the epithelium.

Classification by function of epithelium

Epithelium can have two functions. They can act as a lining (lining epithelia) and/or form secretory structures (glandular epithelium).

Lining Epithelia covers the free surfaces of the body (internal and external) and cavities. Examples of this include epidermis, lining of the gastrointestinal tract and ducts. Their position in contact with the environment gives them great importance in regulating the composition of the body by controlling the movement of materials in and out. Tissues that line the inside of the mouth, the esophagus and part of the rectum are composed of nonkeratinized stratified squamous epithelium. Other surfaces that separate body cavities from the outside environment are lined by simple squamous, columnar, or pseudostratified epithelial cells. Other epithelial cells line the insides of the lungs, the gastrointestinal tract, the reproductive and urinary tracts, and make up the exocrine and endocrine glands. The outer surface of the cornea is covered with fast-growing, easily regenerated epithelial cells. Endothelium (the inner lining of blood vessels, the heart, and lymphatic vessels) is a specialized form of epithelium. Another type, mesothelium, forms the walls of the pericardium, pleurae, and peritoneum.

Glandular epithelium is an invagination or aggregation of epithelium that forms a solid tissue structure. Glandular tissue is formed from glandular epithelium that forms structures from the infolding of epithelium and subsequent growth in the underlying connective tissue.

Glands can be broadly classified into two types : endocrine and exocrine. Endocrine glands secrete their product into the extracellular space where it is rapidly taken up by the blood vascular system. Exocrine glands secrete their products into a duct that then delivers the product to the lumen of an organ or onto the free surface of the epithelium. Exocrine glands are further characterized by the method in which they secrete product. Merocrine secretion involves the exocytosis of secretory vesicles to the lumen of the gland. Holocrine secretion involves the death of an entire cell within a gland which subsequently sloughs off and releases content into the lumen of the gland. Apocrine secretion involves the release of budding vesicles off the epithelial cells into the lumen of the gland.

Apocrine

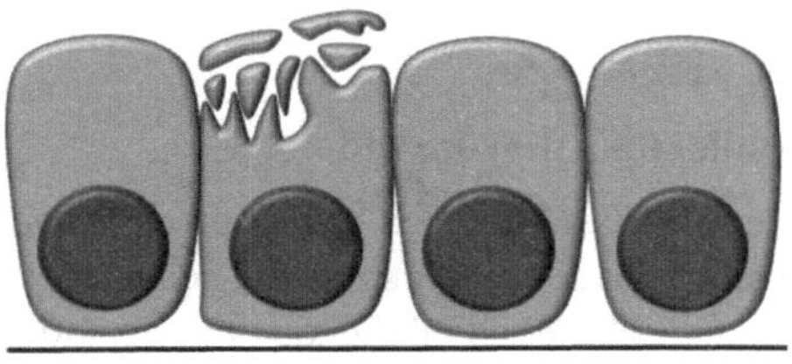

Merocrine

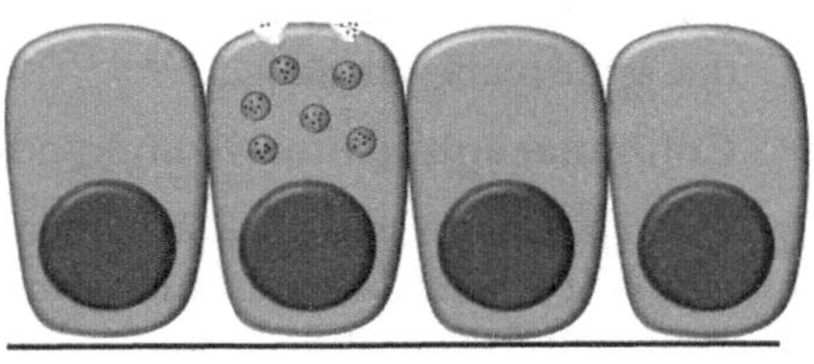

Holocrine

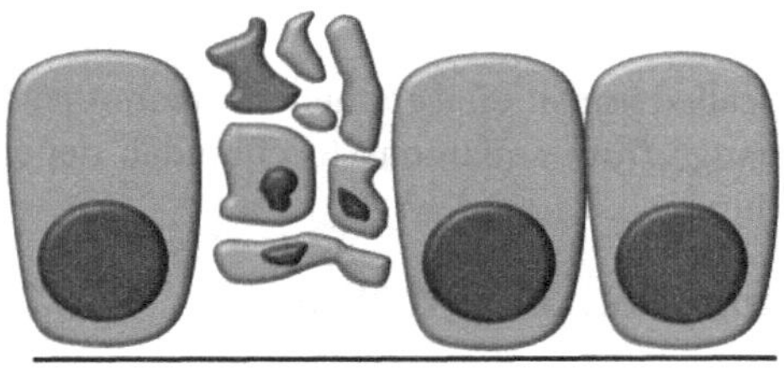

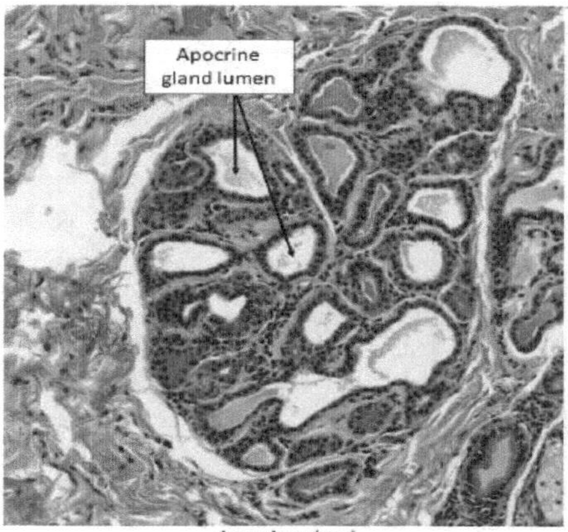

Apocrine gland
Apocrine secretion

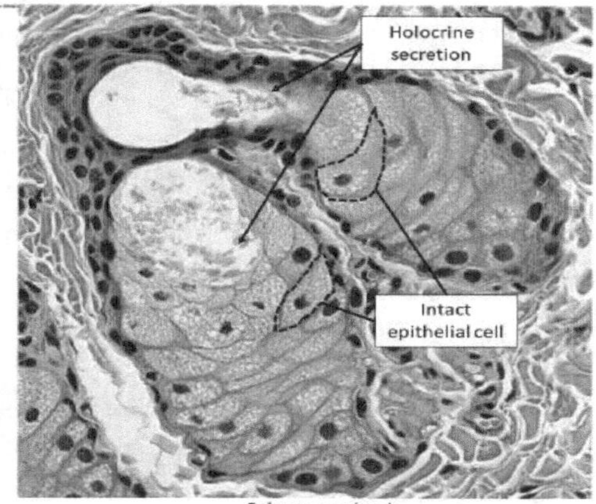

Sebaceous gland
Holocrine secretion

Modifications to epithelium

Microvilli

Microvilli are cell surface modifications which increase surface area of a cell without significantly increasing the size of the cell. Microvilli are often seen in organs where the epithelium plays a primary role in the absorption of molecules. Microvilli are supported by an actin microfilament network. Individual microvilli are difficult to individually resolve by light microscopy and often appear as a band running across the apical aspect of the cell layer.

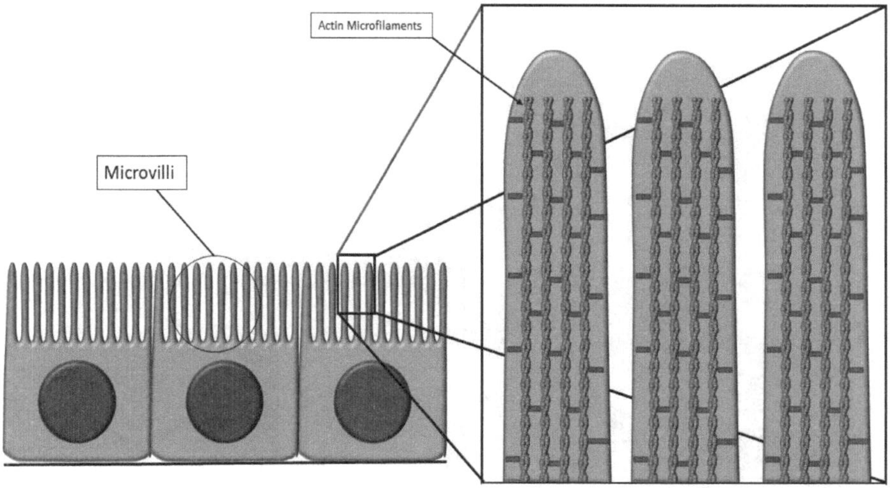

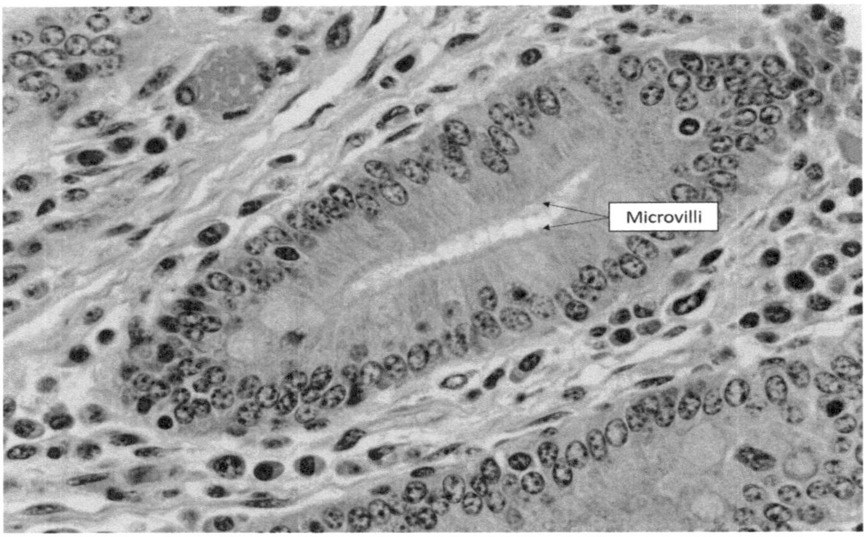

Bovine. Microvilli are present on the apical aspect of the columnar epithelium from the duodenum.

Cilia

Cilia are projections and a type of organelle seen on the apical surface of epithelial cells. In many locations, they beat in a coordinated fashion. This assists in the movement of material over the epithelial surface in a manner parallel with the surface of the epithelium.

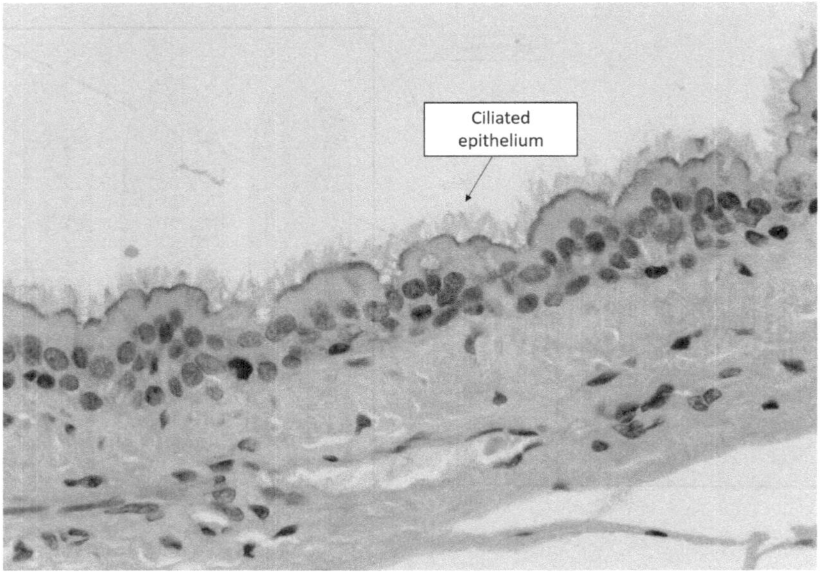

Canine. Pseudostratified epithelium of the trachea with prominent cilia on the apical surface.

Cellular Connections

Cell connections or junctions are especially abundant in epithelial tissues. These structures consist of protein complexes and induce connectivity between adjacent epithelial cells, between a cell and the extracellular matrix. They can contribute to the barrier function of epithelia and control the paracellular transport.

Cell junctions are the contact points between plasma membrane and tissue cells. Some examples of the major types of cell junctions: tight junctions, gap junctions, desmosomes and hemidesmosomes. Tight junctions are transmembrane proteins fused on outer plasma membrane. Gap junctions connect the cytoplasm of two cells and allow for the passage of molecules freely between cells.

Desmosomes and hemidesmosomes allow for strong attachment between cells or to a basement membrane. Desmosomes attach to the microfilaments of cytoskeleton made up of keratin protein. Hemidesmosomes are similar to desmosomes in terms of function, however, they attach the epithelial cell to the basement membrane rather than to an adjacent cell.

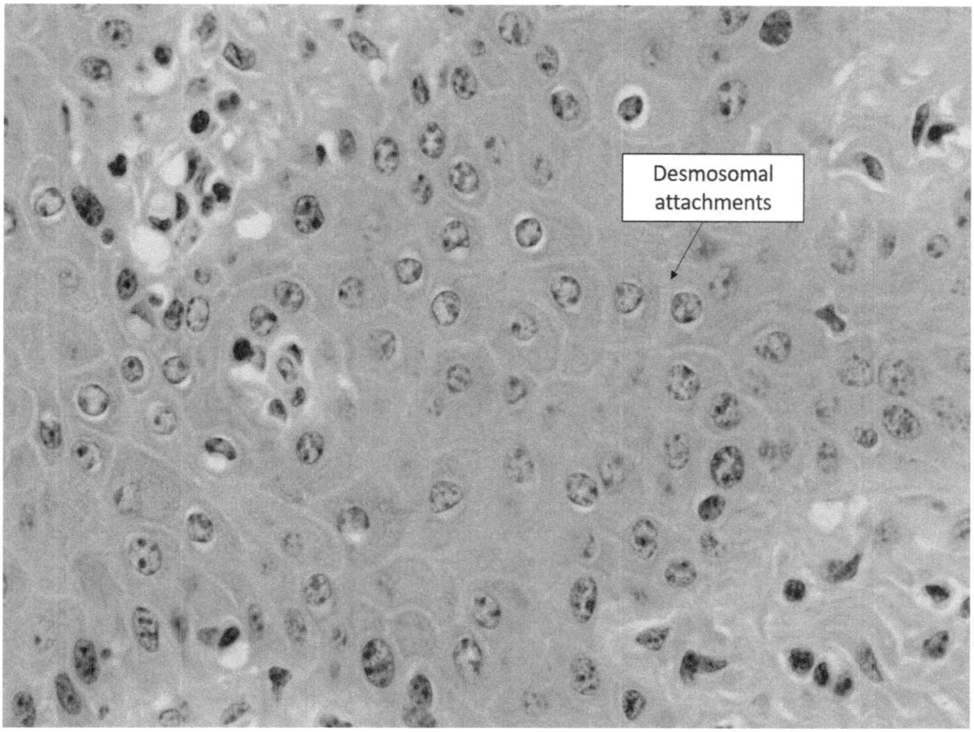

Canine. Prominent desmosomes present between keratinocytes of stratified squamous epithelium.

Chapter-3

Connective tissue

Connective tissues make up the connecting or supporting framework for most of the organs of the body. This chapter will cover connective tissues.

Chapter Learning Objectives

By the end of this chapter, you should be able to do the following:

- Define types and composition of connective tissue of connective tissue
- Define morphologic characteristics of each connective tissue
- Describe the functions of each type of connective tissue

Review Questions

By the end of this chapter, you should be able to answer the following:

- What is the function of connective tissue?
- What are the differences between ground substance, collagen and elastic fibers?
- What are the cellular components of connective tissue?
- What are the differences between embryonic and fibrous connective tissue?
- What is collagen and how is it composed?
- What are the two types of adipose tissue?
- What are the three major types of cartilage and what anatomical locations are they present?

Overview

Connective tissue is a term used to describe the tissue of mesodermal origin that that forms a matrix beneath the epithelial layer and is a connecting or supporting framework for most of the organs of the body.

In contrast to epithelia, connective tissue is sparsely populated by cells and contains an extensive extracellular matrix consisting of protein fibers, glycoproteins, and proteoglycans. The function of this type of tissue is to provide structural and mechanical support for other tissues, and to mediate the exchange of nutrients and waste between the circulation and other tissues. These tissues have two principal components, an extracellular matrix and a variety of support cells. Extracellular matrix is comprised of two non-cellular biological materials, the fiber component and ground substance. In addition, there is a cellular component to connective tissue that is comprised of the cells that produce and maintain the extracellular matrix and that cells that migrate from the associated vasculature to reside in the extracellular matrix.

The fiber component

The different types of connective tissues are specified by the relative content of three distinguishing types of extracellular fibers: **collagenous fibers, elastic fibers, and reticular fibers**.

Collagenous Fibers

The molecules of collagen fibers are oriented to form a lattice in many regions. However in some tissues (tendon, ligaments), the fibers are oriented in dense parallel arrays to provide strength in one direction. These fibers are inelastic, but have great tensile strength. Thus they can be bent without breaking.

While the fibers themselves do not stretch, their lattice-like arrangement can allow tissues containing the fibers to stretch to some extent. Thus, collagen fibers impart both strength and flexibility to tissue.

Collagen is synthesized from the aggregation of precursor molecules, primarily procollagen. Collagen molecules aggregate to larger chains of molecules called fibrils. Fibrils are further organized to create fibers which are visible on light microscopy. Fibers often organized in parallel array forming a bundle.

There are 5 generally accepted types of collagen:

- **Type I**: most common, found in every connective tissue.
- **Type II**: found in hyaline and elastic cartilage and in vitreous body of eye.
- **Type III**: found in reticular fibers, healing wounds, smooth muscle, and fetal skin.
- **Type IV**: found in basal laminae of epithelia
- **Type V**: found in placental basal laminae, tendon, and muscle sheaths.

Elastic Fibers

Elastic fibers are composed primarily of the protein elastin. These fibers are often organized into lamellar (parallel) sheets, as in the walls of arteries. Dense, regular, elastic tissue is an important component in many ligaments and tendons.

Elastic fibers are stretchable because they are normally disorganized – stretching these fibers makes them take on an organized structure.

Reticular Fibers

Reticular fibers are composed of type III collagen. Unlike the thick and coarse collagenous fibers, reticular fibers form a thin reticular network. Such networks are widespread among different tissues and form supporting frameworks in the liver, lymphoid organs, capillary endothelia, and muscle fibers.

Ground Substance

The ground substance is an aqueous gel of glycoproteins and proteoglycans that occupies the space between cellular and fiber elements of the connective tissue. It is characterized by a gel-like consistency. The characteristics of the ground substance determine the permeability of the connective tissue layer to solutes and proteins.

The cellular component

Fibroblasts

Fibroblasts are by far the most common native cell type of connective tissue. The fibroblast synthesizes the collagen and ground substance of the extracellular matrix. These cells make a large amount of protein that they secrete to build the connective tissue layer.

Sometimes the term *fibrocyte* is used to distinguish a small mature cell from a larger metabolically active fibroblast; however, in general convention, the term fibroblast is more common.

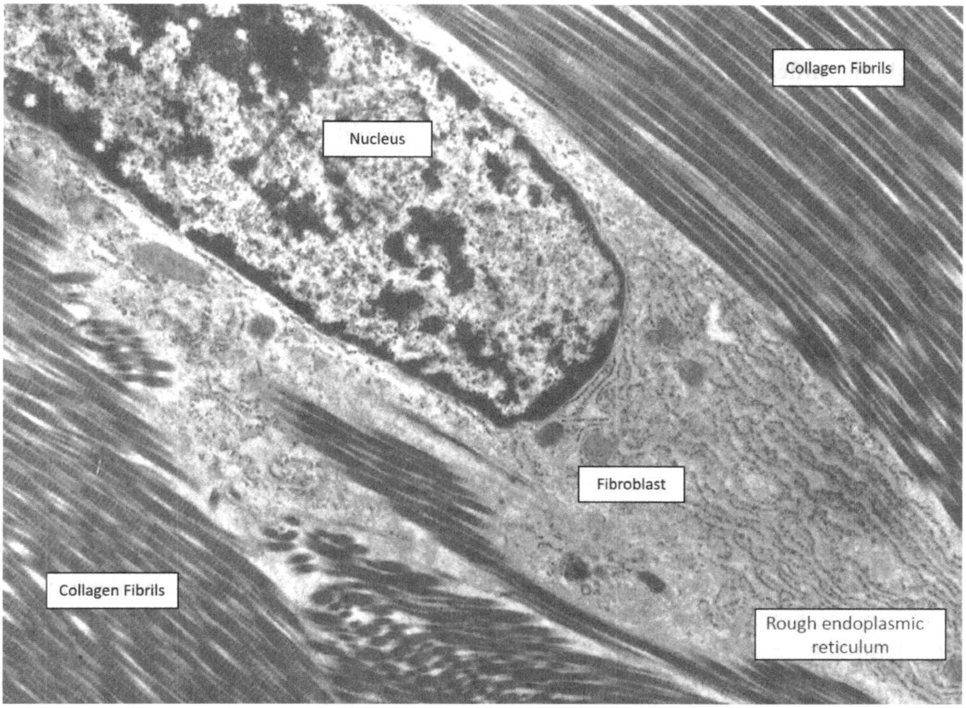

An electron photograph of a fibroblast surrounded by collagen fibrils. Image courtesy of Mamoru Yamaguchi, PhD

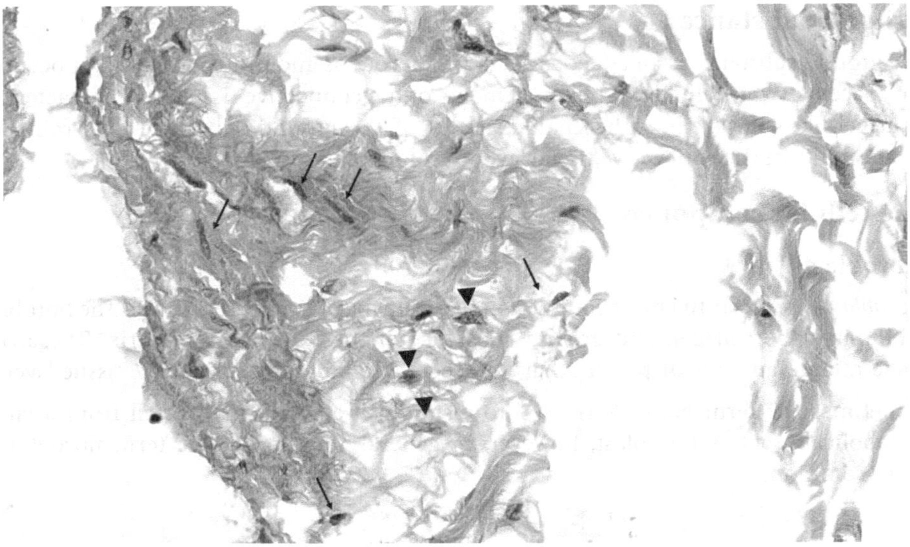

High magnification light photomicrograph depicting fibroblasts in dense irregular connective tissue. The arrows label mature fibroblasts/fibrocytes and the arrowheads depict more immature fibroblasts.

Macrophage

The *macrophage* is the connective tissue representative of the reticuloendothelial, or mononuclear phagocyte, system. This system consists of a number of tissue-specific, mobile, phagocytic cells that descend from monocytes. These include the Kupffer cells of the liver, the alveolar macrophages of the lung, the microglia of the central nervous system. The macrophage is the connective tissue version.

Macrophages phagocytose foreign material in the connective tissue layer and also play an important role as antigen presenting cells.

Other Leukocytes

Other leukocytes are frequently present in small numbers in normal connective tissue. *Lymphocytes* represent an adaptive immune response in tissue. These cells typically have a small amount of cytoplasm, are round and have round nuclei. It cannot be determined from routine hematoxylin and eosin staining as to whether the cells are B lymphocytes or T lymphocytes.

Neutrophils can rarely been seen in connective tissue and are more commonly observed in inflammatory states. Neutrophils look slightly different on histology than on cytology. Unlike in a cytologic preparation, the segmentations of the nuclei are difficult to see and the nucleus frequently appear as joined hyperchromatic round structures in the cytoplasm.

Mast Cells

Mast cells are granulated cells typically found in connective tissue. These cells mediate immune responses to foreign particles. In particular, they release large amounts of histamine and enzymes in response to antigen recognition. This degranulation process is protective when foreign organisms invade the body, but is also the cause of many allergic reactions.

White Adipose Cells

White adipose (fat) cells are specialized for the storage of triglyceride, and occur singly or in small groups scattered throughout the loose connective tissue. When fat cells have accumulated in such abundance that they crowd out or replace cellular and fibrous elements, the accumulation is termed adipose tissue.

These cells usually contain one centrally located vacuole of lipid. The cytoplasm forms a circular ring around this vacuole and is rarely visible. The nuclei of these cells are compressed and displaced to the side. The function of white fat is to serve as an energy source and thermal insulator.

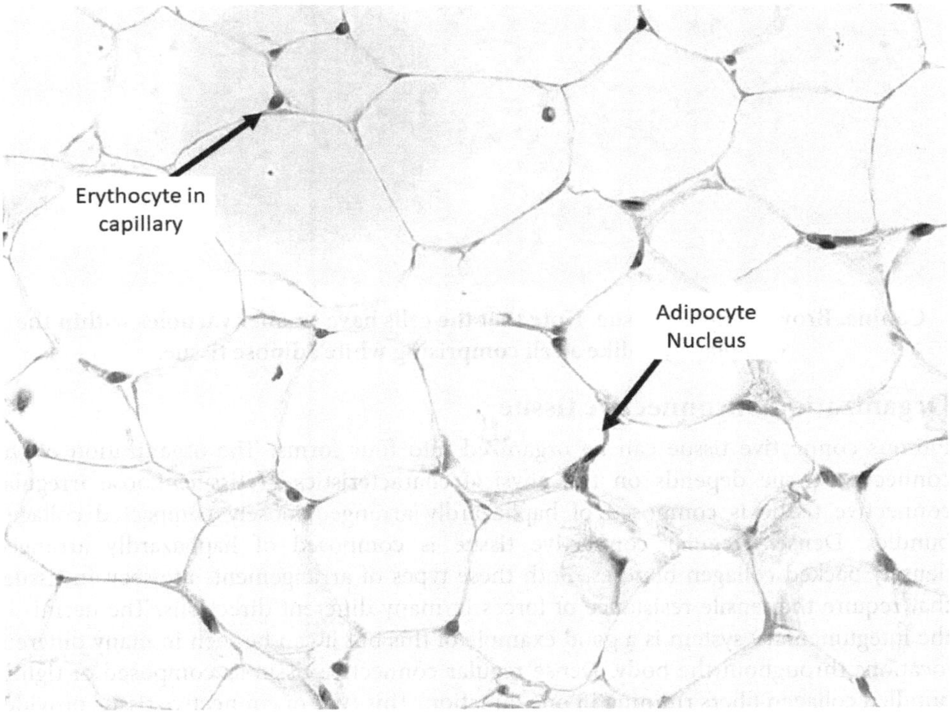

Canine. White adipose tissue. Note that the nuclei are compressed to the side. The actual cytoplasmic lipid content is removed during processing so the cell appears white.

Brown Adipose Cells

Brown adipose cells are highly specialized for temperature regulation. These cells are abundant in newborns and hibernating mammals, but are typically rare in adults. Some species, such as the mouse, have abundant brown adipose stores as adults.

They have numerous, smaller lipid droplets and a large number of mitochondria, whose cytochromes impart the brown color of the tissue.

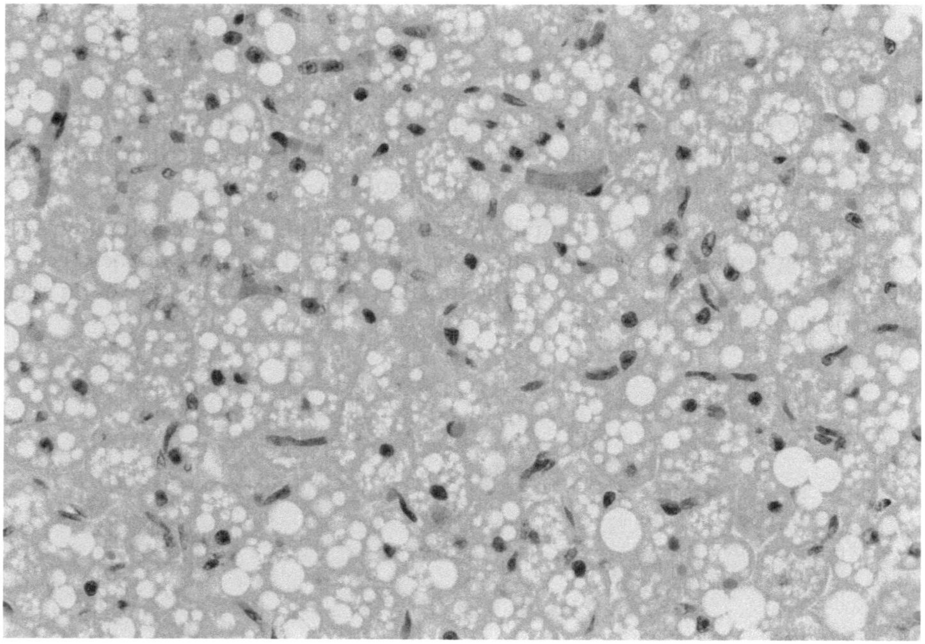

Canine. Brown adipose tissue. Note that the cells have smaller vacuoles within the cytoplasm, unlike a cell comprising white adipose tissue.

Organization of connective tissue

Fibrous connective tissue can be organized into four forms. The organization of the connective tissue depends on the physical characteristics of tissue. Loose irregular connective tissue is composed of haphazardly arranged loosely compacted collagen bundles. Dense irregular connective tissue is composed of haphazardly arranged densely packed collagen bundles. Both these types of arrangements are seen in tissues that require the tensile resistance of forces in many different directions. The dermis of the integumentary system is a good example of this but it can be seen in many different locations throughout the body. Dense regular connective tissue is composed of tightly bundled collagen fibers running in one direction. This type of connective tissue provides excellent tensile strength in one direction. Tendons and ligaments are composed of dense regular connective tissue and are critical in maintaining musculoskeletal structure under the repeated strain of movement. The fourth arrangement of connective tissue is known as embryonic connective tissue and, as the name suggests, is only located in embryonic tissue. This is a very fragile connective tissue composed of thin strands of collagen fibers that contains a large amount of interstitial water content. The fibroblasts in this type of connective tissue frequently has a stellate appearance, meaning that instead of a spindle shape with two pointed ends, the cells may have three pointed projections of cell membrane.

Cartilage

Cartilage is a specialized form of connective tissue produced by differentiated fibroblast-like cells called chondrocytes. It is characterized by a prominent extracellular matrix consisting of various proportions of connective tissue fibers embedded in a gel-like matrix.

Chondrocytes are cells unique to cartilage and are located within lacunae (holes) in the matrix that they have built around themselves. Lacunae become separated from one another as a result of the secretory activity of the chondrocytes.

Three kinds of cartilage are classified according to the abundance of certain fibers and the characteristics of their matrix:

- **Hyaline cartilage** has a matrix composed of type II collagen and chondromucoprotein, a copolymer of chondroitin sulfates A and C with protein. Its high concentration of negatively- charged sulfate groups makes it appear intensely basophilic under H&E staining. This type of cartilage makes up the articular surface in joints.
- **Fibrocartilage** is distinguished by its high content and orderly arrangement of type I collagen fibers. It is typically located in regions where tendons attach to bones, the intervertebral discs, and the symphysis between certain bones.
- **Elastic cartilage** is characterized by the presence of abundant elastic fibers and is quite cellular. It is made up of type II collagen and is located in the pinna of the ear, the turbinates of the nose and the epiglottis.

Chapter-4

Muscle

Muscle is one of the four primary tissue types. There are three types of muscle tissue: **skeletal muscle, cardiac muscle, and smooth muscle**. This chapter will cover muscle.

Chapter Learning Objectives

By the end of this chapter, you should be able to do the following:

- Name the general functions of muscle.
- Classify the three major types of muscle.
- Describe the locations in which each type of muscle is present.
- Described the structural components which differentiate each type of muscle and how it allows each type to function.

Review Questions

By the end of this chapter, you should be able to answer the following:

- What are the three major types of muscle?
- How are these types of muscle different? List the differences.
- What locations are each of these types of muscle present?
- Differentiate the following terms. Myofiber, myofibril, sarcomere.
- List the structural components of a sarcomere. How does a sarcomere function?
- How is cardiac muscle different than skeletal muscle? How are they the same?
- What is the function of smooth muscle?
- What structural components are different in smooth muscle compared to cardiac and skeletal muscle?
- What is myoepithelium?

Overview

Muscle is one of the four primary tissue types. There are three types of muscle tissue: **skeletal muscle, cardiac muscle, and smooth muscle**. The three muscle types have some properties in common. They all exhibit excitability as their plasma membranes change their electrical states (depolarization) and send an electrical signal in a wave-like pattern called an action potential along the entire length of the skeletal muscle cell membrane. Skeletal muscle depends on the nervous system to work properly. This is referred to as voluntary motion. In contrast, the nervous system influences the activity of cardiac and smooth muscle to a lesser degree. On the other hand, both cardiac muscle and smooth muscle can respond to other stimuli, such as catecholamines (epinephrine and norepinephrine) from the blood, hormones and local stimuli.

Myoepithelium and myofibroblasts are specialized types of epithelium and fibroblasts, respectively, that have contractile ability. Myoepithelial cells are often located in and around glands where extracellular product produced by the epithelial cells needs to be moved by contractile action towards the duct system. Myofibroblasts are often seen in healing wounds and assist in maturation and contraction of the granulation tissue (immature fibrous connective tissue) as it matures and contracts.

Skeletal muscle

Function

One of the most important functions of skeletal muscle is to contract and cause movement. Skeletal muscles act not only to produce movement but also to stop movement. Constant small contractions of the skeletal muscle are needed to hold a body upright in any position, even at rest. Muscles also prevent excess movement of the bones and joints, maintaining skeletal stability and preventing skeletal structure damage or deformation. Skeletal muscle is located throughout the body at the openings of internal tracts to control the movement of various substances. This muscle allow voluntary control of specific functions, such as swallowing, urination, and defecation. Skeletal muscles also protect internal organs (particularly the abdominal organs) by acting as an external barrier or shield to external trauma and by supporting the weight of the organs.

Skeletal muscles contribute to the generation of heat, primarily by shivering or intense bursts of movement. Muscle contraction requires energy, and when ATP is broken down, heat is produced. This heat is very noticeable during and following intense active movement such as running. In this case, sustained muscle movement causes body temperature to rise. Similarly, when environmental temperatures are low, random skeletal muscle around the appendicular skeletal produces random skeletal muscle contractions to generate heat (shivering).

Structure

Each skeletal muscle is an organ that consists of various integrated tissues. These tissues include the skeletal muscle fibers, blood vessels, nerve fibers, and connective tissue. Each skeletal muscle has three levels of connective tissue that enclose it. These levels of connective tissue provide structure to the muscle as a whole, and also compartmentalize the muscle fibers within the muscle in to groups. Each large muscle is wrapped in a sheath of dense, irregular connective tissue called the **epimysium**, that holds muscle fascicles in a group and also allows a muscle to contract while maintaining its structural integrity. This is the outermost layer. The epimysium also allows the muscle to move independently from other muscle groups and other types of tissue. Most individual muscles that you learn in gross anatomy are surrounded by epimysium.

Muscle fiber is a term synonymous with muscle cell. The muscle fiber is a cell with multiple peripherally located nuclei that runs the ENTIRE length of the actual muscle. Inside each skeletal muscle, muscle fibers are organized into individual bundles (called **fascicles**), by intermediate layer of connective tissue called the **perimysium**. Inside each fascicle, each muscle fiber is encased in an extremely thin connective tissue layer called the **endomysium**.

Some terminology associated with muscle fibers are rooted in the Greek *sarco*, which means "flesh."

The plasma membrane of muscle fibers is called the **sarcolemma**, the cytoplasm is referred to as **sarcoplasm**, and the specialized smooth endoplasmic reticulum that stores and controls the release of calcium ions (Ca++) is called the **sarcoplasmic reticulum (SR)**. The functional unit of a skeletal muscle fiber is the sarcomere, a highly organized arrangement of the contractile myofilaments **actin** (thin filament) and **myosin** (thick filament), along with other support proteins. Individual sarcomeres are joined together in a long chain called a **myofibril** and attaches to the sarcolemma at its end. Myofibrils are bundled together within the sarcoplasm. In addition, there are invaginations of the sarcolemma into the sarcoplasm that surround the myofibrils. These are referred to as t-tubules. The t-tubules system conveys the action potential from the sarcolemma into the cell in order stimulate the myofibrils.

Distinct lines or bands in the sarcomere are apparent where the myosin and actin filaments overlap as well as where the sarcomere connect to each other consecutively. This is what gives skeletal muscle its striated appearance on histology. In fact, skeletal muscle and cardiac muscle (because cardiac muscle also has sarcomeres) are also known as striated muscle. As the sarcomeres and thus, the myofibrils contract, the entire muscle cell contracts. Because the actin and its troponin-tropomyosin complex (projecting from the Z-discs toward the center of the sarcomere) form strands that are thinner than the myosin, together it is called the **thin filament** of the sarcomere. Because the myosin strands and their multiple heads (projecting from the center of the sarcomere, toward but not all to way to, the Z-discs) have more mass and are thicker, they are called the **thick filaments**. As mention before, in order for a sarcomere to contract, calcium released from the sarcoplasmic reticulum after a stimulatory action potential from the T-tubules is generated. Calcium interacts with molecules on the actin filaments, troponin and tropomyosin, which normally block myosin binding sites on actin. ATP binding to myosin causes release of the myosin head from actin.

Contraction

Contraction of the sarcomere involves several steps. In order to understand these step, you must understand that myosin head has several important characteristics.

- It has ATP-binding sites into which fit molecules of ATP. ATP represents potential energy.
- It has actin-binding sites into which fit molecules of actin.
- It has a "hinge" at the point where it leaves the core of the thick myofilament. This allows the head to swing back and forth, and this motion is what causes muscle contraction.

During contraction the following steps occur:

1. The activated myosin head attaches to the actin binding site; contraction refers to the activation of myosin's cross bridges – the sites that generate the force
2. The phosphate group, P, on the myosin head falls off
3. The myosin head then produces a power stroke which results in the sliding of the filaments forcing the thin filaments to move towards each other. The ADP on the myosin also falls off.
4. Once the power stroke is complete, ATP again attaches to the myosin head causing the head to detach from the actin site and return to its original position.
5. ATP attached to the myosin head is split by ATPase causing the myosin heads to be activated again; the cycle can then be repeated over and over again as long as calcium and ATP are present. Multiple cycles of this process are often necessary for full muscle contraction.

Cardiac muscle

Cardiac muscle tissue is only found in the heart, specifically the myocardium. Highly coordinated contractions of cardiac muscle pump blood into the vessels of the circulatory system such as the aorta and pulmonary artery. Cardiac muscle is striated and organized into sarcomeres, possessing the same organization of myofilaments as skeletal muscle. However, cardiac muscle fibers are shorter than skeletal muscle fibers and usually contain only one nucleus, which is located in the central region of the cell. Cardiac muscle fibers also contain many mitochondria. This is necessary as large amounts of ATP are required by the cells and this is produced primarily through aerobic metabolism.

Cardiac muscle fibers cells also are extensively branched and are connected to one another at their ends by **intercalated discs**. An **intercalated disc** allows the cardiac muscle cells to contract in a coordinated fashion so that the heart can work as a pump. Intercalated discs are part of the sarcolemma and contain two structures important in cardiac muscle contraction: gap junctions and desmosomes. Gap junction form channels between adjacent cardiac muscle fibers that allow the depolarizing action potential to move between one cardiac muscle cell to the next. It allows the quick transmission of action potentials and the coordinated contraction of the entire heart. The remainder of the intercalated disc is composed of desmosomes that were discussed in the epithelium chapter. **Desmosomes** anchor the ends of cardiac muscle fibers together so the cells do not pull apart during individual fiber contraction.

Smooth Muscle

Smooth muscle (named as such because the cells do not have striations) is present in the walls of organs that contain a lumen like the urinary bladder, uterus, stomach, intestines, and in the walls of large vascular tubes, such as the arteries and veins of the circulatory system. The tracts of the respiratory, urinary, and reproductive systems contain smooth muscle. Smooth muscle is also present in the eye, where it functions to change the size of the pupil and in the skin where it causes hair to stand erect in response to low temperatures or fear. Smooth muscle is referred to as an involuntary muscle since is not under voluntary control. In certain locations, such as the walls of visceral organs, stretching the muscle can trigger its contraction (the stretch-relaxation response).

Smooth muscle contraction relies on the presence of Ca++ ions similar to skeletal and cardiac muscle. However, smooth muscle fibers are much smaller in all dimensions than skeletal muscle cells. T-tubules are not required to reach the interior of the cell and therefore not necessary to transmit an action potential deep into the fiber. Smooth muscle fibers have a limited calcium-storing sarcoplasmic reticulum but have calcium channels in the sarcolemma that open during the action potential along the sarcolemma. The influx of extracellular Ca++ ions, which diffuse into the sarcoplasm to reach a protein called calmodulin. The calcium-calmodulin complex activates myosin light chain kinase, an enzyme that phosphorylates myosin and allows it to bind to the actin network within the cell.

Muscle contraction continues until ATP-dependent calcium pumps actively transport Ca++ ions back into the sarcoplasmic reticulum and out of the sarcoplasm. A small amount of calcium remains in the sarcoplasm to maintain muscle tone and keep the cells contracted to a small degree. This is important in certain organs and around blood vessels. Because most smooth muscles must function for long periods without rest, their power output is relatively low, but contractions can continue without using large amounts of energy.

Chapter-5

Bone, Cartilage, and Joints

This chapter will cover bone, cartilage, and joints.

Chapter Learning Objectives

By the end of this chapter, you should be able to do the following:

- ❖ Define the components of bone on the microscopic level when bone is developing
- ❖ and when an animal is skeletally mature.
- ❖ Describe the function of bone
- ❖ Described the methods in which bone is formed.
- ❖ Classify the types of cartilage present in an animal.
- ❖ Describe the structure of each type of cartilage and list the locations where each is present.
- ❖ Describe the structure of tendons, ligaments and synovium.

Review Questions

By the end of this chapter, you should be able to answer the following:

- ❖ What are the major organic and inorganic constituents of bone?
- ❖ What are the structure and function of the following cell types:
- ❖ Osteoblast
- ❖ Osteocyte
- ❖ Osteoclast
- ❖ What are the anatomic structures that facilitates communication between osteocytes?
- ❖ Explain the basic process of osteoclastic resorption, naming one important stimulus for resorption?
- ❖ What is the difference between osteoid and bone?

What is the difference between woven and lamellar bone?

- ❖ What is the difference between cancellous and compact bone?
- ❖ What is an osteon?
- ❖ What does the term modeling mean in term of bony tissue?

- ❖ What does the term remodeling mean in term of bony tissue? Compare and contrast
- ❖ remodeling and modeling
- ❖ List the fundamental processes in remodeling, in the proper sequence
- ❖ List two instances where bone remodeling plays an important role in NORMAL bone function
- ❖ What does the process of intramembranous ossification involve? Where does it occur?
- ❖ What does the process of endochondral ossification involve?
- ❖ Name sites where this occurs
- ❖ Explain the difference between a primary and secondary ossification center
- ❖ Define the articular epiphyseal complex
- ❖ List the three processes which are essential for endochondral ossification to occur
- ❖ normally
- ❖ What is the physis and the significance of this finding in a histologic section?
- ❖ What are the three types of cartilage and what locations can they be found?
- ❖ What are the predominant components of cartilage ground substance?
- ❖ What is the structure and function of the following:
- ❖ Chondroblast
- ❖ Chondrocyte
- ❖ Perichondrium
- ❖ Sharpey's fiber
- ❖ Intervertebral disc
- ❖ Synovial membrane

Overview

Bone is a dynamic organ that functions to provide support, protection, and also act as a reservoir of mineral salts and growth factors. The histologic structure of bone is highly specialized to perform these functions.

Bone microanatomy

Bone is primarily composed of extracellular matrix, with organic and inorganic components. The organic component is largely type i collagen and ground substance (collectively this is called osteoid). Non collagenous proteins that contribute to osteoid include growth factors, 'glues', and proteins that help to initiate mineralization. The inorganic matrix is calcium hydroxyapatite (this also contains phosphorus), which adds rigidity. Due to this rigidity, bone can only undergo appositional growth (layers on top of each other), not interstitial growth (growth from within the matrix). Routine histologic sections of bone are generally demineralized, and mineral content becomes impossible to evaluate.

The arrangement of collagen fibers in bone can be visualized under a polarizing microscope. Where fibers are arranged into dense parallel arrays, the bone is referred to as 'lamellar'. All normal bone in an adult animal is lamellar. In the cortex, there are outer (periosteal) circumferential lamellae and inner (endosteal) lamellae. Rapidly deposited osteoid (such as occurs during growth and injury) has randomly arranged collagen fibers and is referred to as 'woven'.

Osteoblasts are mesenchymal cells that contribute to bone production in two phases. First, they synthesize and secrete a layer of organic materials (primarily tropocollagen) called the osteoid seam. Secondly, osteoblasts contribute to mineralization of the osteoid seam (at which point the matrix becomes bone). Osteoblasts are plump, angular cells that can be seen lining bone surfaces. With time, osteoblasts become embedded within bone and become osteocytes.

Osteocytes are smaller and elongate, and are contained within small cavities called lacunae. They have long cytoplasmic process that run within canaliculi, connecting to other osteocytes (these cell processes are very hard to identify in routine histologic sections). Changes in interstitial fluid movement and electric potential (piezoelectric forces) can be detected by the osteocytes, which can then transmit a signal to osteoblasts. By this mechanism, osteocytes play an important role in detecting local changes in the microenvironment (stress, microfractures, micronutrients) and co- ordinating adaptive responses. Osteocytes may also play a role in day to day calcium homeostasis by 'fine-tuning' the amount of calcium that is released/removed from the circulating calcium pool.

In a section of long bone, bone can anatomically be divided into compact bone and trabecular (also called cancellous or spongy) bone. Compact bone is the dense bone that forms the cortex. Trabecular bone forms multi-directional, anastomosing struts within the marrow cavity, and are interspersed by adipose and hematopoietic tissue. The arrangement of trabecular bone generally reflects loading patterns on the bone. Covering bone surfaces, there are two distinct envelopes: the periosteum and endosteum. The periosteum covers the outermost aspect of the cortex. There are two layers of the periosteum, the outer (fibrous) layer and the inner osteogenic (cambium) layer. In a normal adult bone, the osteogenic layer may not be identifiable. The osteogenic layer of the periosteum is easily identified and 'active' during periods of growth and following injury. The endosteum lines the innermost margin of the cortex, bone trabeculae, and osteonal canals.

Osteoclasts are hematopoietic derived, multinucleated cells that resorb bone. Physiologic stimuli that induce osteoclastic resorption generally operate through the osteoblast (the osteoblast secretes a molecule called ODF that binds RANK on osteoclasts, causing osteoclast activation). The osteoblast initially releases collagenases that resorb the non-mineralized lamina limitans. The peripheral margin of the osteoclast can then bind to the exposed mineralized bone, forming a tight seal. Carbonic acid secreted across a brush border of the osteoclast resorbs the mineral and the matrix is dissolved by secreted proteinases. A Howship's lacuna is the scalloped cavity left after an osteoclast resorbs a segment of bone.

Osteons are the functional unit of bone. They can be compared to reinforcing rods embedded in concrete. An osteon is a long, cylindrical structure with a central osteonal canal (also called Haversian canal) containing nerves, blood vessels and osteogenic cells. This central canal is surrounded by densely arranged concentric lamellae of bone, that are approximately 2-3μm apart. Primary osteons initially form at the periosteal surface. Secondary osteons form after primary osteons. To form, secondary osteons must have a pre-existing tunnel allowing osteoclasts to enlarge the tunnel. Osteoblasts then lay down bone in a centripetal direction, thus forming a secondary osteon. The reversal line marks the point at which osteoclasis stops and osteogenesis begins. Interstitial bone can be seen between osteons, and represents remnants of primary/ secondary osteons and the circumferential lamellae.

Bone function

The different constituents of bone function together as a unit, to facilitate normal growth, adapt to changing stress, repair microinjury, regulate mineral homeostasis, and respond to injury.

Changes in bone size and shape are called modeling. Modeling is the way in which bone responds to loading. It is exemplified by normal growth, as the developing bone shapes to a contour that allows it to withstand loading. Bone will also model during periods of abnormal use. The modeling process occurs by Wolff's Law, which states that bone shape and orientation will adapt to use. The response is formation at sites of compression and resorption at sites of tension. When modeling occurs slowly, the deposited bone is lamellar; where modeling occurs rapidly the deposited bone is woven. Modeling differs remodeling as bone can directly begin forming or resorbing (not the same activation- resorption-formation sequence as outlined below).

During remodeling, bone is removed from one area and replaced in another. Remodeling plays a crucial role in normal cortical growth, as woven bone is replaced by osteonal bone. This process is also important in repairing 'wear and tear' (microcracks), and replacing old bone with new bone. The sequence of remodeling is always the same: activation, osteoclastic resorption, reversal, formation. Collectively, osteoblasts and osteoclasts combine to form the 'basic metabolic unit' of bone. Activation begins as osteoblasts are stimulated to contract and remove the lamina limitans (activation factors include PTH and others). Reversal is the inhibition of resorption and initiation of formation. During formation, the space left by osteoclasts becomes filled by osteoid deposited by osteoblasts. An osteonal remodeling unit is constantly changing. The osteon has a central cavity shaped like an ice-cream cone with osteoclasts at the top leading edge (called the 'cutting cone'). Further away from the top, the cavity left by the cutting cone becomes progressively filled by new bone deposited in concentric lamellae (giving the remodeling unit a coned appearance in longitudinal section). There are progressively more centripetal lamellae as sections are further from the site of osteoclastic resorption. When osteoclastic activity ceases, the top of the remodeling unit becomes filled and becomes a mature secondary osteon.

Bone growth

Bone can grow by two mechanisms: membranous ossification and endochondral ossification. Membranous ossification occurs primarily in flat bones (ie. bones of the skull, mandible), while endochondral ossification is seen in long bones (ie. appendicular skeleton).

Membranous ossification

Membranous ossification begins as condensation of primitive mesenchyme, and cells differentiate into an osteoblastic phenotype. These cells form bone directly that is laid in apposition, on top of the bone that has just formed. Importantly, there is no cartilage intermediate. Membranous ossification is also the way in which bones (including long bones) grow in width. It therefore contributes a significant portion to the diaphyseal cortex of long bones.

Endochondral ossification

Endochondral ossification significantly differs from membranous ossification. A cartilage model that becomes almost entirely replaced by bone precedes the formation of the actual bone. Increases in bone length are primarily the result of endochondral ossification, as the cartilage model can grow interstitially (from within the matrix). In the developing fetus, primitive mesenchyme condenses and these cells differentiate into a chondroblastic phenotype. The chondroblastic population proliferates to form a cartilage model that approximately mimics the final bone contour. Because this model is initially composed entirely of cartilage (considerably less rigid than bone), the developing bone can grow interstitially. Initially, the perichondrium that surrounds the cartilage model becomes invaded by blood vessels, forming periosteal collar around the diaphyseal region. With this periosteum, there are osteogenic precursors that form an bony sleeve around the primary center of ossification. This primary center of ossification begins in the middle of a long bone as blood vessels invade into the cartilage model. With time, vessels then invade the periphery of the bone, establishing secondary centers of ossification. The primary and secondary centers of ossification progressively expand with endochondral ossification occurring at the edges of an ossification center. Once the ossification centers get to a certain point, there becomes a thin line of growing cartilage between the primary and secondary ossification center. This is known as the physis ('growth plate') and its primary function is to lengthen the bone. Endochondral ossification that occurs at the edge of a secondary ossification center, immediately beneath an articular surface, is called the articular-epiphyseal complex (AEC). Histologically the process of endochondral ossification is almost identical. Endochondral ossification in any given ossification center occurs at different rates at different loci. This allows the bone to form its shape and spatial alignment. Once endochondral ossification finishes at the AEC, the subchondral bone condenses to form an end plate.

The process of endochondral ossification can be seen in histologic sections of a growth plate (physis), where multiple distinct layers can be identified. These include the zone of resting chondrocytes, zone of proliferation, zone of hypertrophy, zone of mineralization, and zone of ossification. Chondrocytes within the zone of proliferation and hypertrophy are arranged into linear columns interspersed by chondroid matrix. For endochondral ossification to occur normally, three things must happen in a very orderly fashion. These are mineralization of the cartilage matrix surrounding the chondrocyte, death of the chondrocyte (mineralization stops transfer of nutrients causing the cell to die), and vascular ingrowth. Vessels grow between spicules of mineralized matrix. This process is particularly important as the new vessels bring osteogenic cells (osteoblasts) that can then deposit osteoid on top of the mineralized cartilage. The area containing mineralized cartilage spicules and early ossification is referred to as the primary spongiosa. In the secondary spongiosa, the newly formed bony trabeculae become modeled into mature trabeculae that reflect tension and compression on the bone.

Cartilage microanatomy

Three distinct types of cartilage exist, and their location depends on the required function.

Hyaline cartilage is the most common type of cartilage in the body. It covers the surfaces of articular joints and is the cartilage present in physes. Hyaline cartilage is primarily composed of water, present as a viscoelastic hydrated gel, reinforced by type II collagen. Water is bound by proteoglycan, the largest of which is a molecule called aggregin. Collagen fibers within the cartilage provide tensile strength and contribute to shock absorption. The deepest aspect of cartilage is mineralized (the junction between mineralized and unmineralized cartilage is the "tidemark"), which anchors cartilage to the bone. Turnover of proteoglycan in cartilage is high and controlled by chondrocytes. This is in contrast to collagen, which has low turnover. Adult chondrocytes have limited replicative potential.

Fibrocartilage is stronger and more robust than hyaline cartilage due to abundant collagen that is arranged in bundles. These bundles are interspersed by islands of cartilage. This cartilage can be found in menisci and the annulus fibrosus of intervertebral discs. Fibrocartilage is sparsely vascularized and has limited response to injury. Elastic cartilage is similar to hyaline cartilage but contains many elastin fibers. These fibers are best demonstrated with special stains. Elastic cartilage is strong but very flexible; elastic cartilage is best demonstrated in the pinna.

Hyaline cartilage is separated from the surrounding tissue by perichondrium. Like periosteum, it has an outer fibrous layer and an inner cellular layer (but has chondrogenic rather than osteogenic potential). This cellular layer is only prominent during periods of growth or injury. The perichondrium is lost from the surface of weight-bearing articular hyaline cartilage (although a thin rim of perichondrium is maintained around the articular edge and is continuous with the periosteum). Perichondrium also covers the periphery of physes ('perichondrial ring'), supporting lateral expansion of the physis by appositional growth.

Chondroblasts are only found within the perichondrium. Morphologically, chondroblasts are small and oval with hyperchromatic nuclei and clear space surrounding the cell (this is an artifact). Chrondroblasts lie in small lacunae, and are often found in groups. Chondroblasts undergo hypertrophy into chondrocytes, taking an oval shape. Chondrocytes are the most frequent cell within hyaline cartilage. Ultrastructurally, chondrocytes have extensive cellular processes, but these are difficult to appreciate by light microscopy. The function of chondrocytes is to produce and maintain the surrounding chondroid matrix. Cartilage can undergo both appositional growth (cells/matrix are deposited on top of each other from the perichondrium) and interstitial growth (growth from within the cartilage model). Hyaline cartilage is also avascular and must be nourished by synovial fluid. This process is facilitated during periods of heavy loading.

Tendons and joints

Lining the innermost aspect of the joint capsule is lined by synovium. Synovium is thought to contain three distinct cell populations: phagocytic macrophages, antigen presenting cells, and fibroblast like cells that produce glycosaminoglycans. Synovial fluid is a filtrate of plasma enriched with proteoglycans. The synovial membrane is normally supported by a loose fibrovascular stroma admixed with adipose. The outermost layer, the joint capsule, is composed of dense collagen. Tendons and ligaments look histologically similar and are also composed of very dense collagen. The collagenous fibers in tendons insert directly into bone, and these are called 'Sharpey's fibers'. In doing so, the tendon is strongly anchored to the underlying bone. Sharpey's fibers can easily be visualized under a polarizing microscope.

Chapter-6

Cardiovascular System

This chapter will cover the cardiovascular system.

Chapter Learning Objectives

By the end of this chapter, you should be able to do the following:

- ❖ Describe the different components of the circulatory system.
- ❖ Identify the tunica intima, tunica media, and tunica adventitia, naming the major
- ❖ components of each.
- ❖ Describe how the vascular tunics differ at different levels of the vasculature (ie. artery vs. vein)
- ❖ Explain the structure and function of endothelial cells.
- ❖ Describe the differences in structure between arteries, arterioles, metarterioles, capillaries,
- ❖ venules, veins and lymphatics.
- ❖ Explain the differences between continuous, fenestrated and sinusoidal capillaries.
- ❖ Describe the structures that comprise the heart.
- ❖ List the microscopic differences between skeletal and cardiac muscle

Review Questions

By the end of this chapter, you should be able to answer the following:

- ❖ What are functions and the components that comprise the following:
- ❖ Elastic artery
- ❖ Muscular artery
- ❖ Capillary
- ❖ Sinusoid
- ❖ Postcapillary venule

Vein

- ❖ Lymphatic vessel

- Vasa vasorum and nervi vasorum
- Arteriovenous anastomosis
- Glomus body
- What is the role of endothelium in hemostasis?
- What are the differences between a continuous, fenestrated, and discontinuous capillary. List the anatomic locations where each can be found?
- How lymph is formed?
- How does a lymphatic vessel differs from a continuous capillary?
- What are functions and the components that comprise the following:
- Pericardium
- Endocardium
- Myocardium
- Intercalated disc
- Valve
- Purkinje cell
- Cardiomyocyte
- Mesothelial cell

Overview

The vasculature can broadly be divided into blood and lymphatic circulations. The blood circulatory system serves to transport oxygen and nourish tissues, remove carbon dioxide as waste products, regulate temperature, distribute humoral factors and deliver immune cells. While blood vessels move blood away from and towards the heart (arteries and veins, respectively), flow in lymphatics moves only towards the heart eventually reaching the cranial vena cava by way of the thoracic duct. In both the blood and lymphatic circulation, the anatomy directly reflects the functional requirements of any given vessel.

Blood circulation

The blood circulation has a systemic circuit and a pulmonary circuit. The latter operates at considerably lower pressure to accommodate oxygenation of blood in the pulmonary capillaries. The microscopic anatomy of both systems is similar, although systemic vessels tend to have thicker walls to accommodate higher pressures.

The blood circulation is composed arteries, arterioles, microcirculation (capillaries and sinusoids), venules, and veins. Arteries serve to conduct high-pressure blood to an organ. These transition into arterioles and metaarterioles that distribute and regulate pressure to a vast network of delicate capillaries. Following exchange at the level of the microvasculature, blood collected by venules is returned to the heart by veins. Generally speaking, blood vessels closer to the heart have wider diameter and thicker walls; arteries have thicker walls and smaller lumens than veins.

Vascular tunics

With the exception of capillaries and sinusoids, all larger vessels have the same three basic structural elements (tunics). These are the tunica intima (inner or luminal layer), tunica media (middle layer), and tunica adventitia (outer layer). Depending on the level of the vasculature, there are marked differences in the tunic width.

Endothelial cells are arguably the most important component of the tunica intima and form the primary barrier between blood and tissue. They are highly specialized cell type and cover the luminal surface of the tunica intima. Endothelial cells have a number of critical functions. During a resting phase, endothelial cells have important anti-clotting functions. They have anticoagulant effects (expression of heparin like molecules, production of thrombomodulin, tissue factor pathway inhibitor), inhibit platelet aggregation (ADPase, prostacyclin), and promote fibrinolysis (synthesis of tissue type plasminogen activator). Under certain pathologic states, endothelium contributes to a prothrombotic state by activating/aggregating platelets (ie. platelet activating factor, von Willebrand's factor), inhibiting fibrinolysis (plasminogen activator inhibitors), and producing procoagulant factors (ie. tissue factor). Other functions include:

1. Synthesis of collagen and proteoglycan for basement membrane maintenance (connective tissue interface)
2. Modulation of blood flow by secretion and metabolism of vasoactive mediators (ie. nitric oxide, endothelin, angiotensin I)
3. Degradation of catecholamines
4. Expression of surface molecules that facilitate adhesion, rolling, and diapedesis of inflammatory cells
5. Production of growth hormones (ie. FGF, PDGF)
6. Production of molecules that mediate the acute inflammatory reaction (ie. IL-1, 6, 8)
7. Antigen presentation

Endothelial cells can be identified by their location. They are the flattened, fusiform cells lining the lumen of vessels. Von Willebrand's factor can be contained in Weibel-Palade bodies that might be apparent under electron microscopy. Other ultrastructural features include cytoplasmic vesicles that reflect phagocytic and pinocytotic activity. Highly specialized endothelial modifications can be present depending on the organ and location (see discussion on capillary structure and function). Supporting the vascular endothelium, and forming the remainder of the tunica intima, is a basement membrane and a subendothelial layer. The subendothelial layer contains collagen, elastic fibers and proteoglycan; smooth muscle cells, fibroblasts and myofibroblasts are also present. Collagen, contained within the subendothelium, is a particularly important stimulus for platelet activation and adhesion. Fibronectin helps to stabilize cell-cell and cell-substrate interactions.

The tunica media is the middle portion of the vessel wall. It contains extracellular matrix (elastic laminae, ground substance), and circumferentially arranged smooth muscle cells. The internal and external elastic laminae delineate the inner and outer limits of the tunica media, respectively. These elastic laminae become harder to identify in closer proximity to the microcirculation. In histologic sections of arteries, arteries contract and the internal elastic lamina has extensive tortuous folds. In arteries, the tunica media is extremely thick and is the primary constituent of the vessel wall. The venous supply, which operates at considerably lower pressure, has an extremely thin tunica media. Unlike arteries, the internal elastic lamina is only prominent in medium and large caliber veins.

The tunica adventitia is the outermost layer of the vessel wall. It is composed of a loose layer of connective tissue, containing vasa vasorum, a vascular bed designed to perfuse the vessel wall itself. The tunica adventitia is particularly prominent in large caliber veins (such as the caudal vena cava), where it may also contain bundles of loosely arranged smooth muscle responsible for maintaining venous tone. The adventitia of large veins is richly innervated by nerves and also contains lymphatic vessels.

Arteries and arterioles

Arteries can be classified as elastic or muscular. Elastic arteries lie closer to the heart and their primary function is to conduct blood to different areas of the body. This blood flows at high pressure, with marked changes in blood velocity. To accommodate this, elastic arteries have thick walls containing many, concentrically arranged, fenestrated elastin laminae. These are interspersed by smooth muscle cells and ground substance rich in proteoglycan. As elastic vessels become progressively smaller, the pulsatile nature of blood flow is more a result of elastic recoil rather than the heat beat itself.

Moving away from the heart, arteries contain progressively less elastin and become muscular arteries. This can be an abrupt or gradual transition. Muscular arteries have a distributive function, serving to take blood to specific organs. Branching and increased number of arteries means that the total volume is increased and flow velocity is reduced. Muscular arteries contain a thick wall of helically arranged smooth muscle cells, with lesser quantities of elastin.

Arterioles are less than approximately 100 μm wide in the contracted state. Regulation of blood flow to the capillary bed happens at the level of the metaarteriole. Precapillary sphincters can also limit flow to the capillaries that cannot tolerate pressures normally encountered in the arterial system.

Blood flow to the capillary bed can be yet further regulated by arteriovenous anastomoses, which allow blood to bypass regional capillaries. Anastomoses arising in close proximity, and surrounded by a fibrous capsule, are referred to as a glomus bodies. The importance of arteriovenous anastomoses is best demonstrated in erectile reproductive tissue.

Capillaries and sinusoids

Capillaries have the smallest diameter of the vascular system. Their anatomy is such that it easily facilitates interchange of nutrients/waste between blood and organ. Capillaries are tubules less than 10 μm wide (ie. not much larger than a red blood cell). Capillaries are often collapsed in routine histologic sections and it can be difficult to appreciate this extensive, anastomosing network of small caliber vessels. Structurally, capillaries are essentially formed of endothelium supported by a basal lamina. The latter is selectively porous to small molecules approximately <70,000 MW. Loosely distributed around capillaries are pericytes that are directly apposed to the basal lamina and are capable of transdifferentiating into different cell types (ie. fibroblasts, smooth muscle cells). The most ubiquitous microcirculatory vessel in the body is the continuous capillary, with endothelial cells linked by tight junctions and intercellular gaps of approximately 1 nm. Examples of continuous capillaries are found in the lungs and brain.

Capillary structure can be yet further modified to reflect the functional needs of an organ. Fenestrated capillaries contain small cytoplasmic gaps and lie on an uninterrupted basal lamina. These vessels are more permeable than continuous capillaries, allowing easy transfer between the blood and interstitial fluid. Fenestrated capillaries are common in endocrine organs and the intestine. Porous capillaries are demonstrated in the glomerulus and allow formation of the ultrafiltrate that ultimately becomes urine.

Sinusoids are slightly larger than capillaries and are found in the adrenal glands, liver, spleen, and bone marrow. Sinusoids have large gaps between endothelial cells; the basal lamina is either discontinuous or absent. As a result, sinusoids are considerably more permeable and are capable of considerable rapid nutrient exchange. Endothelial cells lining hepatic sinusoids are interspersed by macrophages, which have an important phagocytic function.

Venules and veins

The structure of venules and veins is essentially similar structure to arterioles and arteries. In contrast to arteries, the tunica media is considerably thinner and the tunica adventitia may form the thickest wall component. Post capillary venules have a particularly important function in inflammatory responses, contributing to fluid leakage and leukocyte diapedesis. This function is not readily apparent in routine histologic sections. All veins and venules contain semilunar valves that prevent retrograde flow.

Lymphatic circulation

Lymph is fluid that drains from the extracellular space of tissues. This fluid normally forms as capillary fluid passes out of the vessels, and is composed of water, electrolytes, and lesser amounts of plasma proteins. Lymph vessels return this fluid to the circulating blood. It is a one "circuit system" until the two interconnected circulatory system comprised by the arteries and veins.

Lymph flows at extremely low pressure and filters through lymph nodes which act as the sentinels for immune stimulation. Because of the low pressure, the walls of lymphatics are exceedingly thin and are readily collapsible (often not apparent in routine histologic sections). The basement membrane is largely absent and there are no pericytes. As a result, lymphatic endothelium is considerably more permeable than continuous blood capillary endothelium. Anchoring filaments attach the endothelial cells to surrounding connective tissue. Valves in lymphatic vessels prevent retrograde flow. The major conduit for lymph in the body is the thoracic duct, which drains into the left brachiocephalic vein (canine; species variations exist). Elevations in central venous pressure have potential to disrupt lymphatic flow, resulting in chylothorax. Tissues that notably lack lymphatics include the central nervous system, cartilage, bone, bone marrow, placenta, cornea, and teeth.

Overview of the heart

The heart and associated structures are broadly broken into 6 functionally and structurally distinct compartments: the pericardium/epicardium, endocardium, myocardium, valves, fibrous skeleton, and conduction system.

Pericardium

The pericardium has fibrous and serous components. The fibrous component forms the outermost layer of the pericardial sac and is composed of dense collagenous tissue. The serous pericardium lines the pericardial cavity, and is covered by mesothelial cells which secrete the pericardial fluid. Quiescent mesothelial cells are flattened under routine light microscopy. Following injury, these cells can become cuboidal or even columnar.

The parietal (outer-most) layer forms the pericardial sac which is readily apparent grossly. This is continuous with the visceral layer (synonymous with the epicardium) which is a thin layer covering the surface of the heart. Beneath the visceral pericardium, there is often a thin layer of adipose containing vessels and lymphatics. While the parietal pericardium is capable of adapting to changes in size of the heart, it has limited ability to expand rapidly (such as following acute pericardial effusion). Chronic distension of the pericardial sac is better tolerated as collagen fibers in the pericardium presumably remodel.

Endocardium

The endocardium is the inner most lining of the heart and is the cardiac equivalent of the tunica intima. It is lined by endothelial cells and supported by a small amount of fibrous stroma. The endocardium is continuous with the atrioventricular valves and also covers chordae tendinae, papillary muscles, and chordae tendinae. The endocardium may contain a small amount of adipose, lymphatic vessels, and nerves. Purkinje fibers, important in electrical transmission, are also contained within the endocardium.

Myocardium

Like skeletal muscle, cardiac muscle is striated with many of the same ultrastructural features, including Z-lines, A bands, and I bands (described in Chapter 4 Muscle). Key differences from skeletal muscle are:

1. Branching of cardiomyocytes
2. Prominent connective tissue
3. Cardiomyocytes have centrally located nuclei with a perinuclear halo
4. Transversely orientated intercalated discs

A unique feature of atrial cardiomyocytes is the presence of cytoplasmic granules containing atrial natriuretic factor (anf). Secretion of anf counteracts aldosterone, and induces natriuresis, vasodilation, and suppression of the renin-angiotensin-aldosterone system. the intercalated discs are especially important for cardiac muscle function, and are junctional complexes between terminal Z lines. The intercalated disc contains desmosomes, gap junctions, and fasciae adherentes. Cardiomyocytes are particularly rich in mitochondria and have large quantities of myoglobin. Ventricular cardiomyocytes are arranged into outer, middle and inner layers that form a convoluted helix around the heart. These layers are difficult to appreciate in routine sections. Cardiac myocytes also form a thin sleeve of tissue that extend from the left atrium into the walls of pulmonary veins. These cells are thought to be an important focus of ectopic activity in certain disease states (such as in atrial fibrillation).

Valves

Valves are covered by endothelium which has a similar function to vascular endothelium elsewhere in the body. Different layers of the tricuspid and mitral valves include the atrialis (on the atrial side), spongiosa, fibrosa, and ventricularis (on the ventricular side). Beyond the proximal ½ of the valve, the valve is avascular and these different layers are not readily appreciable. The valve stroma is composed of a mixture of collagen, elastic fibers, and proteoglycans. Valve interstitial cells (VICs) are the primary stromal cell and are responsible for homeostasis of the extracellular matrix. VICs are also activated in valvulopathies, and play an important role in progression of diseases such as myxomatous valvular degeneration ('endocardiosis').

Conduction System

Heart rate is determined by a collection of sinoatrial nodal cells ('pacemaker cells') located near the confluence of the cranial vena cava and right atrium, in the subepicardial myocardium. While these cells are essentially modified cardiomyocytes, they are slightly smaller and interspersed by dense fibrous connective tissue. Sinoatrial node cells are often found in close association to the sinus node artery. They are extensively modified by autonomic ganglia which are located in the adjacent epicardium. The atrioventricular node is microscopically similar to the sinoatrial node, and can be found slightly cranial (anterior) to the coronary sinus, and immediately dorsal to the tricuspid annulus. Electrical impulses continue to be conducted through Purkinje fibers which branch at the basal aspect of the interventricular septum, and send ventricular ramifications in the subendocardium. Purkinje cells are considerably larger than regular cardiomyocytes and stain lightly eosinophilic. Electrical conduction is facilitated by desmosomes and gap junctions. Intercalated discs are not present between Purkinje cells.

Chapter-7

Integument

The integumentary system, or skin, represents the interface between the internal body and the external environment. This chapter will cover the integument.

Chapter Learning Objectives

By the end of this chapter, you should be able to do the following:

- Describe the histologic appearance, organization and function of each of the four layers of the epidermis: stratum basale, stratum spinosum, stratum granulosum, and stratum corneum.
- Describe the histologic appearance and physiologic roles of the dermal adnexa, including
- sebaceous glands, apocrine (epitrichial) glands, and hair.
- Describe the physyiologic roles of melanin and melanocytes.
- Describe and illustrate the normal histologic anatomy of the carnivore digit/claw.
- Describe and illustrate the normal histologic anatomy of the hoof and apply this anatomy to the concept of "laminitis."
- Describe the source of the three layers of hoof wall stratum corneum.
- Compare and contrast the name, topography, and functional roles of specialized/ modified
- sebaceous and apocrine glands.

Review Questions

By the end of this chapter, you should be able to answer the following:

- Which layer of the epidermis is primarily responsible for proliferation?
- What molecular structures are responsible for cell-cell adhesion in the epidermis?
- Name two protective roles of melanin in the epidermis.
- Name two protective roles of hair/feathers.

Sebaceous adenitis in dogs is associated with immune-mediated destruction of sebaceous glands. What clinical symptoms do dogs with sebaceous adenitis present with (Hint: think of the normal function of sebum, and how the absence of sebaceous glands would impact this)?

- Damage to the coronary band would be expected to impact which layers of the hoof wall?
- Identification of rotation of P3 (by radiographs) is a classic feature of severe laminitis in horses. Why does P3 rotation occur secondary to laminitis (Hint: recall the anatomic structures of this region and their functions)?

Overview

The integumentary system, or skin, represents the interface between the internal body and the external environment. As such, the primary role of skin is **protection**. This includes protection from physical insults, including UV light, as well as providing a barrier to pathogens. In addition, the skin is crucial to an animal's **thermoregulation**, **sensing** of its environment, and **energy and metabolic** functions.

The skin is composed of three primary components: *Epidermis, Dermis, and Subcutaneous tissue* (also known as the *hypodermis* or *subcutis*).

Structure of the epidermis

The epidermis is a multilayered epithelial structure composed of keratinocytes that produce the structural protein keratin. Histologically, the epidermis is composed primarily of four tightly-adherent layers of squamous epithelium organized into distinct vertical zones by stages of differentiation.

The epidermis is reminiscent of a brick wall, with the keratinocytes representing bricks, and intercellular matrix representing the mortar. From innermost to outermost the layers are: stratum basale, stratum spinosum, stratum granulosum, and stratum corneum.

Stratum basale

The deepest/innermost layer of the epidermis is the stratum basale. Histologically, the stratum basale is a single layer of cuboidal keratinocytes that directly abut and attach to the dermis. The two primary functions of the stratum basale are 1) proliferation and 2) attachment of the epidermis to the dermis.

Cells of the stratum basale essentially represent germinal cells ("brick generators") responsible for the generation of all cells of the epidermis. As such, these cells are the most mitotically active keratinocytes (apparent in histologic sections) and are the least differentiated. Stratum basale keratinocytes divide and then "push up" towards the surface, differentiating into cells of the stratum spinosum.

Stratum spinosum

Histologically, the stratum spinosum is located directly above (superficial to) the stratum basale. Cells of the stratum spinosum have prominent cell-to-cell junctions, termed desmosomes, that appear as spiky membrane projections on histology. For this reason, these are referred to as "prickle cells". Desmosomes are protein complexes that firmly attach keratinocytes to their surrounding cells (*intercellular*, between cells), thereby providing significant structural integrity. Tonofilaments are *intracellular* (within cells) protein complexes that anchor desmosomes to the cell membranes.

The thickness of the stratum spinosum can vary dramatically depending on the anatomic location on an animal. For example, the stratum spinosum of canine inguinal skin is 1-2 cell layers thick, whereas the stratum spinosum of the canine footpad is often >20 cell layers thick! The reason for this is clear when the primary function of skin is considered: protection.

Stratum granulosum

The stratum granulosum, where present, sits above the stratum spinosum and directly below the stratum corneum. Histologically, the stratum granulosum is a thin layer of keratinocytes containing dense, basophilic (blue/purple) cytoplasmic structures called keratohyalin granules. Keratohyalin granules contain components responsible for keratinization, including fibrous proteins (keratohyalin) and a lipid-rich secretory product.

It should be noted that the stratum granulosum is not apparent in all regions of the skin, particularly in regions of thin skin (e.g. canine inguinal and axillary skin).

Stratum corneum

The stratum corneum is the variably thick (10-20 layers) outermost layer of the skin. Histologically, cells of the stratum corneum, or corneocytes, are flattened eosinophilic keratinocytes that lack nuclei (anuclear). Corneocytes, often considered "dead", retain some metabolic and signaling functions despite having no organelles or nuclei.

Keratinocytes of the stratum corneum have a thickened, insoluble, hydrophobic, cell membrane and contain abundant keratin. In addition, the corneocytes are surrounded by a hydrophobic phospholipid secretory product (produced by stratum granulosum). Together, these combine to provide the hydrophobic and structural properties that serve to protect the epidermis from physical insults as well as water loss. Desmosomal attachment become more sparse in superficial layers, resulting in sloughing of keratinocytes, a process known as desquamation.

Physiology of the epidermis

The epidermis is essentially a layering of differentially mature keratinocytes, going from least differentiated (stratum basale) to most differentiated (stratum corneum). The rate and extent of differentiation may change according to physiologic demand, whether due to normal, everyday need (e.g. digital pads) or in response to a disease process (e.g. bacterial dermatitis). In this way, the skin responds to external, and sometimes internal, stimuli in an exceptionally dynamic manner.

Another important point to consider is the energy needs of the integumentary system. The constant proliferation and generation of new keratinocytes is necessary to balance the constant loss of superficial keratinocytes (desquamation). As a result, the skin has an immense nutritional requirement. In the event of suboptimal nutrition, the skin and hair coat may be altered, reflecting this systemic nutritional deficiency. In fact, dogs with chronic malnutrition often have epidermal atrophy and other integumentary changes. Additionally, vitamin and mineral deficiencies (e.g. Vitamin A, zinc, copper, et al.) manifest with integumentary changes including hyperkeratosis, defective hair coat quality, and predisposition to secondary bacterial infections of the skin. Therefore, the skin often reflects both the nutritional and health status of an animal, and is therefore a clinically important aspect of the physical exam.

Additional cells of the epidermis

Although keratinocytes comprise the vast majority of epidermal cells, there are a number of accessory cells present in the normal epidermis that provide important contributions to the function of the skin.

Melanocytes

Melanocytes are professional melanin synthesizing cells located primarily within the basilar (deeper) regions of the epidermis, and are histologically recognized as poorly staining "clear cells". Variation in epidermal melanin content is responsible for the color variety in most domestic species. The density of melanocytes within the epidermis is relatively uniform between individual animals of a species. Pigmentation is therefore largely dependent on the *amount* of melanin produced by melanocytes. Melanocytes have complex dendritic processes that lie adjacent to keratinocytes and facilitate the transfer of melanin to keratinocytes. Although melanocytes are the producers of melanin, keratinocytes often contain the greatest amount of melanin, via transfer from melanocytes.

Melanogenesis, the process of melanin generation, occurs within specialized, melanocyte-specific organelles called melanosomes. Production of melanin can vary based on anatomic location (e.g. oral mucosa pigmentation in some dogs) and/or in response to physiologic need (e.g. chronically sun- exposed skin). Tyrosine is an amino acid required for melanogenesis. Within melanosomes, tyrosine is converted into intermediate compounds (e.g. dihydroxyphenylalanine, DOPA) by tyrosinase. Tyrosinase is a copper-containing enzyme. For this reason, copper deficiency in sheep and cattle results in a characteristic loss of hair coat pigmentation.

Melanin serves unique protective roles in the epidermis, including the ability to both scatter and reflect ultraviolet (UV) rays, and as a scavenger of potentially damaging free radicals. Melanization is also a dynamic process, and increased pigmentation (hyperpigmentation) or reduced pigmentation (hypopigmentation) may be associated with various disease cutaneous processes.

Langerhans cells

Langerhans cells are dendritic cells sparsely located within the stratum basale. Although they are functionally different from melanocytes, they also appear histologically as "clear cells". As dendritic cells, the primary function of Langerhans cells is to take up antigens (antigen = molecules that induce an immune response) and present them to and activate cells of the immune system, such as T cells. In this role, Langerhans cells serve as "sentinels" at the interface of the external environment and the immune system. For example, in a bacterial skin infection, Langerhans cells may take up fragments of the bacteria (bacterial antigens), migrate to the regional lymph node, and present these bacterial antigens to T cells present in the lymph node. In doing so, the Langerhans cell bridges aspects of the innate (skin) and adaptive (T cells) immune responses.

Structure of the dermis

The dermis serves as the structural support foundation of the brick wall that is the epidermis. The dermis is composed of a meshwork of collagen-rich fibrous connective tissue. Collagen is the primary structural protein of the dermis. Elastin, an elastic protein, is present in lesser amounts but contributes to flexibility of the dermis. The structural proteins are interspersed with a gel-like ground substance composed of extracellular fluid and glycoaminoglycans. In this way, the dermis is similar to a sponge where dynamic compression is possible.

The dermis is divided into two regions: the superficial, or *papillary dermis*, and the more substantial *reticular dermis*. The functions of the dermis, in addition to providing structural and tensile strength, include thermoregulation (vasculature), support for adnexal structures, a focus of immune responses, and storage of fluid, electrolytes and nutrients.

Dermal cell populations

Dermal collagen, elastin and ground substance is produced by dermal fibroblasts, narrow spindloid mesenchymal cells uniformly but sparsely distributed throughout the dermis. Dermal fibroblasts not only generate the structural proteins of the dermis, but also play a major role in dermal inflammation and wound healing. In addition, the dermis contains low numbers of resident immune cells, including mast cells, macrophages, dendritic cells, and T cells. These immune cells are critical in the development of both innate and adaptive immune responses in the skin.

Vasculature

The vascular supply for the epidermis and dermal adnexal structures is located within the dermis. The vasculature is divided into three distinct layers: subcutaneous/deep dermal, cutaneous plexus, and papillary plexus. The largest blood supply, the subcutaneous/deep dermal supply, is composed of arteries and veins that branch superficially into the middermis to form the cutaneous plexus. The cutaneous plexus is responsible for supplying blood to the adnexal structures of the dermis, including follicles and glands. Finally, the cutaneous plexus branches superficially to form a fine network of capillaries and venules called the papillary plexus, which lies directly subjacent to and supplies blood to the epidermis.

Blood flow to the skin is dynamic, and factors such as environmental temperature, systemic disease and/or localized inflammation can readily induce shunting of blood to or from the skin.

Dermal adnexa - follicles

The dermal adnexa, or appendages, are accessory structures that include hair and feather follicles as well as secretory glands.

Hair and feather follicles

Hair and feathers serve as an adjunct barrier to the skin, providing a thermal barrier, physical barrier, and a photo-protective barrier. Hair and feathers are produced from follicles. Follicles are tubular structures formed by invaginations of epidermis into the dermis. Follicles serve to generate, attach, and support hair or feather shafts, and also contribute to epidermal re-epithelialization and repair following injury.

Hair follicles have three distinct zones; from deep to superficial they are: bulb, isthmus, and infundibulum. Hair growth begins at the deepest portion of the follicle, the bulb. A specialized structure, the dermal papilla, lies adjacent to the bulb and induces growth of the hair follicle and hair production. Hairs/feathers are essentially rolled cylinders of keratinocytes. On section, hair/feathers have three distinct zones; from outermost to innermost, they are: cuticle, cortex, and medulla.

Hair and feather growth occurs in a cyclical pattern of proliferation progressing through distinct stages of development which include anagen (growth phase), catagen, and telogen (resting phase). Terminally, the hair is exfoliated (shed). Growth is influenced by numerous factors that include anatomic location, hormones, daylight, ambient temperature, nutrition, and various growth factors.

Hair *shafts* are categorized into primary and secondary hairs. Primary hairs are larger and also referred to as "guard hairs", whereas secondary hairs are smaller, and comprise the "undercoat". Species such as horses only have primary hairs; dogs and cats have both primary and secondary hairs. This, in part, accounts for the different texture of their coats. Further, hair *follicles* are categorized as simple and compound. These categories are based on the number of hair *shafts* that exit a single infundibulum. Horses and cattle only have simple follicles (one hair shaft emerges from one infundibulum), whereas dogs and cats have both simple follicles interspersed with compound follicles (multiple hair shafts emerge from one infundibulum). In sheep, wool is produced as compound follicles.

Arrector pili muscle

The arrector pili muscle is a band-like bundle of smooth muscle that attaches to the hair follicle and dermis. The primary role of the arrector pili muscle is pilo-erection (hair erection). Pilo-erection serves to establish a layer of air between the surface of the hair/fur and the skin, serving as insulation.

In addition, pilo-erection has a major role in animal behavior for domestic species such as the cat and dog.

Tactile hairs (vibrissae, sinus hairs, whiskers)

Tactile hairs are modified hair follicles surrounded by a blood-filled sinus and highly innervated. Because of this arrangement, motion of the hair transmits through the blood-filled sinus to the nerve, transducing physical stimulus to sensation.

Dermal adnexa - sebaceous glands

Sebaceous glands are composed of clusters of pale-staining, highly vacuolated epithelial cells (sebocytes) that are located adjacent to follicles. Sebaceous glands are responsible for producing sebum, a lipid-containing compound with moisturizing and antimicrobial properties. Sebocytes secrete sebum in a manner called "holocrine secretion", in which the sebocytes rupture, thereby releasing their cytoplasmic contents (sebum) into the sebaceous duct. The sebaceous duct and the sebum empties into the isthmus portion of the hair follicle, coating the hair shaft and surrounding epidermis in sebum.

Specialized sebaceous glands by location

Meibomian gland – Eyelid

The Meibomian glands are sebaceous glands that open through a common duct onto the margin of the eyelid. Histologically, the sebocytes of the Meibomian gland appear identical to those of sebaceous glands in the skin, but are organized into a much larger, single gland. The Meibomian gland's lipid- rich secretion (meibum) contributes to the tear film.

Glands of Zeis – Eyelid

Glands of Zeis are specialized sebaceous glands that are associated with eyelashes, specialized hair follicles located at the margins of the eyelid. The Glands of Zeis are histologically identical to cutaneous sebaceous glands located elsewhere.

Hepatoid (perianal) glands – Perianal, parapreputial, and tail skin

Hepatoid glands are modified sebaceous glands in dogs, so named due to their morphologic similarity to hepatocytes. These glands are primarily located in the perianal skin (perineum), as well as along the base of the tail and prepuce. They are more developed in male dogs. Histologically, they appear as discrete lobules composed of an outer layer of small basaloid cells surrounding larger, polygonal cells with abundant eosinophilic (pink) cytoplasm (resembling hepatocytes).

These glands are of clinical importance as hepatoid gland adenomas (benign tumors) are common. Tumors of hepatoid glands should be distinguished from the tumors of the apocrine glands of the anal sac (discussed below).

Dermal adnexa - epitrichial (apocrine)

Glands and eccrine glands

Epitrichial glands and eccrine glands are coiled tubular glands. Epitrichial glands secrete via apocrine secretion in which small portions of membrane surrounding cytoplasmic contents are actively broken off and secreted into the gland lumen. Eccrine glands secrete via merocrine secretion, in which secretion is by exocytosis, and secrete directly onto the surface of the skin. These apocrine glands empty into the hair follicle (epitrichial glands) or onto the skin surface (eccrine glands).

Apocrine glands are widely distributed in dogs and cats, always associated with haired skin. Eccrine sweat glands, however, are present primarily in the non-haired skin of the footpads and nose.

Specialized apocrine glands by location

Anal sac glands – Perineum

The anal sacs are paired saccules that are located deep to the perineum and immediately ventrolateral to the rectum in carnivores (cats and dogs). The anal sacs have ducts that empty directly into the terminal rectum. Anal sacs are lined by a thin layer of keratinizing stratified squamous epithelium and surrounded by skeletal muscle. These structures serve to receive and excrete the secretions of specialized glands, called apocrine glands of the anal sac (anal sac glands) that line the wall of the anal sac. Although histologically identical to apocrine glands of the skin, the anal sac glands are functionally distinct, and produce an odiferous secretion that likely serves in olfactory communication (territory, etc.).

The anal sacs and their glands are of significant clinical importance. Apocrine gland adenocarcinomas of the anal sac are common and highly malignant tumors in dogs developing from these anal sac glands. Additionally, inflammation and impaction of the anal sacs are common clinical diseases.

Glands of Moll – Eyelash

Glands of Moll are specialized apocrine glands that are associated with eyelashes. The Glands of Moll are histologically identical to apocrine glands of the skin.

Ceruminous glands – External ear canal

The ceruminous glands are modified apocrine glands throughout the skin of the external ear canal.

These glands are histologically identical to apocrine glands of the skin elsewhere. The ceruminous glands contribute to cerumen, a waxy product composed of the combined secretions of ceruminous glands and sebaceous glands of the external ear canal.

Hypodermis (subcutis, subcutaneous tissue)

The subcutis is the deepest layer of the integument, sandwiched between the dermis and skeletal muscle. The hypodermis serves as physical protection, as well as an energy reserve and source of insulation and thermal regulation. It is composed primarily of adipose tissue, the amount of which varies based on species, anatomic site, and nutritional status. Adipose tissue is composed of clusters or sheets of lipid-filled cells, called adipocytes. On histology, the lipid in adipocytes appears empty, as the fat-soluble lipid is washed out during processing.

Specialized anatomic sites

Histology of the Digit

The anatomy of the digits of domestic species are varied, and some species, such as horses and ruminants, have highly specialized digits worth additional consideration.

Carnivore digit

In carnivores, the ventrum of the digit is called the digital pad. The digital pad is the toughest and thickest skin of the body. The digital cushion is an accumulation of abundant fibroadipose tissue (similar to the subcutis) that is also rich in sensory nerves and glands.

The claw (nail) of carnivores is composed of specialized epidermis and dermis that are firmly affixed to the underlying third phalanx (P3). Growth of the claw occurs outward from the epidermis adjacent to the nail fold. The externally apparent outer layer of the claw is composed of abundant layers of compact stratum corneum (anuclear cornified keratinocytes). This stratum corneum is the portion of the claw that is "trimmed".

Beneath the epidermis is the highly vascular and well-innervated dermis, or "quick". The deep dermis fuses with the periosteum of the ungual process of the distal P3. In this manner, the dermis anchors the claw to the digit.

Equine hoof

The equine hoof is an anatomically complex structure. The hoof itself can be compared to the claw of the carnivore, as it represents abundant epidermis with underlying dermis attached to P3. The dorsal and lateral hoof is called the wall, whereas the ventral aspect of the hoof is called the sole. The sole is anatomically similar to that of the digital cushion of carnivores.

The hoof wall is composed of three distinct layers of stratum corneum. The outer two layers, the stratum externum and stratum medium, are generated by and grow downward from the epidermis of the coronary band (or coronet, similar to the cuticle of a fingernail). The stratum externum, or periople, is the thin, outermost layer of the hoof wall. The stratum medium, or tubular horn, is the middle and thickest layer of the hoof wall. This layer also may contain pigment.

The innermost layer of the hoof wall is the stratum internum. This layer is produced by the epithelium of the laminar epidermis and is fused to the stratum medium, forming a relatively contiguous hoof wall. The laminar epidermis sits on top of the laminar dermis, also referred to as the corium. The corium is firmly affixed to P3, similar to that of the carnivore digit.

One of the most critical and specialized structures of the hoof are the laminae (singular: lamina). To understand the anatomy of the hoof, it is beneficial to understand the physiologic roles of the hoof.

The forces exerted on the hoof by animals such as the horse are significant. The hoof is required to maintain its structural integrity despite these enormous forces (consider the forces exerted on the hooves of a galloping horse)! Hoof laminae are anatomic adaptations that facilitate this structural integrity. Laminae are interdigitating finger-like projections of dermis (dermal laminae) and epidermis (epidermal laminae). The interdigitation of these laminae generates friction at the interface when force is applied. What's more, these interdigitations are extensive, and there are both primary laminae as well as smaller projections, called secondary laminae. Secondary laminae greatly increase the surface area of interdigitation between the dermal laminae and epidermal laminae. An increase in surface area of contact between these structures corresponds to increased frictional surface and increased resistance to separation, allowing for optimal structural integrity.

Although structural integrity at the interface of dermal and epidermal laminae is maintained in normal animals, separation at this interface can occur, and is referred to as laminitis. Separation of laminae results in loss of structural integrity and the characteristic radiographic finding of rotation of the third phalanx (P3).

Chapter-8

Gastrointestinal system

The gastrointestinal system of domestic species exemplifies the level of anatomic diversity the domestic species commonly encountered in veterinary medicine. Despite this diversity, the gastrointestinal system of all species is unified through the primary goal of breaking down ingested nutrients for absorption and utilization.

This chapter will cover the gastrointestinal system.

Chapter Learning Objectives

By the end of this chapter, you should be able to do the following:

- Describe the histologic layers and subdivisions of the tongue and oral cavity.
- List the histologic and functional differences between glandular and nonglandular gastric mucosa.
- Describe the histologic and functional characteristics of the following gastrointestinal cell types:
- Goblet cells
- Parietal cells
- Chief cells
- Paneth cells
- Describe the interspecies variations of the stomach (gastric) compartment(s), and how this
- correlates with function.
- Describe and identify the following structures of the tubular digestive tract:
- Mucosa
- Muscularis mucosa
- Submucosa
- Muscularis externa (tunica muscularis)
- Serosa
- Submucosal plexus and myenteric plexus

Identify and describe small intestinal microvilli and villi, and understand the functional significance of these epithelial modifications.

- ❖ Describe the histologic location and functional role(s) of the Crypts of Lieberkuhn (intestinal crypts).
- ❖ Describe the histologic features that distinguish duodenum, jejunum, and ileum (i.e., how can you determine which segment you are examining based on histology?)
- ❖ Describe the location(s) and function of the gastrointestinal-associated lymphoid tissue (GALT).

Review Questions

By the end of this chapter, you should be able to answer the following:

- ❖ What are the functional roles of gastric Chief cells and Parietal cells, and how do they contribute to the process of digestion?
- ❖ What structures and cells normally reside within the lamina propria of the small and large intestine?
- ❖ Consider the normal function of intestinal villi and microvilli. Infection with enteric pathogens, such as transmissible gastroenteritis virus of pigs (porcine coronavirus) results in damage to the intestinal enterocytes and villous collapse (atrophy) and fusion. What are the expected clinical consequences of this damage to enterocytes and villi and why (pathogenesis)?
- ❖ Parvoviruses, such as canine parvovirus-1 and feline panleukopenia virus, specifically infect cells of the Crypts of Lieberkuhn within the GI tract, resulting in death of cells within the intestinal crypts. What are the clinical consequences of these infections? What are the histologic changes expected following infection of a puppy with canine parvovirus-1 (i.e. if CPV-1 kills cells of the crypts, what consequences does this have to the intestinal mucosa?)

Overview

The gastrointestinal system of domestic species exemplifies the anatomic diversity encountered in veterinary medicine. Despite this diversity, the gastrointestinal system of all species is unified through the primary goal of breaking down ingested nutrients for absorption and utilization. This process begins with prehension and ingestion, and follows with digestion, absorption, and waste elimination. Knowledge of the microscopic anatomy of the gastrointestinal tract goes hand-in-hand with understanding the functionality of this system. Additionally, knowledge of the histology will greatly aid the veterinarian in understanding the processes, infectious and otherwise, that mediate clinical disease in these species.

Oral cavity

The major structures of the oral cavity are the lips, teeth, tongue, oral mucosa and salivary glands. The primary function of the oral cavity is in prehending, masticating, and moistening of food. The latter two functions prepare food for deglutition (swallowing).

The oral mucosa (labial and gingival) offers protection during mastication and deglutition, and is composed of stratified squamous epithelium that may be keratinized. Below the mucosa is a dense layer of vascular collagen-rich stroma, the lamina propria.

Mobile regions of the oral cavity, such as the soft palate, may also contain skeletal muscle that aids in mastication and deglutition.

Tongue

The tongue is a muscular organ that serves two major roles: mechanical manipulation of ingesta and sensing (taste). The tongue is covered by typical oral mucosa (stratified squamous epithelium) overlying a lamina propria and large amounts of organized skeletal muscle (discussed below).

The dorsal surface of the tongue is decorated by specialized structures called papillae. These structures serve diverse functions, ranging from facilitating taste (e.g. circumvallate) to grooming (e.g. filiform papillae in cats). There are four distinct papillae: filiform papillae, fungiform papillae, foliate papillae, and circumvallate papillae. The filiform papillae generally predominate in the rostral portion of the tongue. Species such as cats have well-developed filiform papillae which contribute to the "sandpaper-like" feel of their tongues. Circumvallate and fungiform papillae are partially lined by taste buds, specialized chemoreceptors that transmit the sensation of taste. These papillae are often located caudally on the dorsal surface of the tongue and are bilaterally symmetric in distribution. They may be prominent structures and should be not be misinterpreted as pathologic lesions.

The core of the tongue is composed of abundant skeletal muscle that is arranged in three distinct layers of alternating orientations/planes (vertical, horizontal, and longitudinal). This robust musculature facilitates the extreme mobility of the tongue.

Salivary Glands

Salivary glands, which include parotid, submaxillary, sublingual, zygomatic (carnivores) and accessory glands, contribute to the production of saliva. Saliva is a complex watery secretion that acts to lubricate, moisten and digest ingesta and is composed of a mixture of mucus, electrolytes, antibodies and digestive enzymes (e.g. amylase).

Structurally, salivary glands are composed of lobules of exocrine tubuloacinar glands that secrete into a duct. Secretion is mediated, in part, by contraction of myoepithelial cells that surround acini. Salivary glands are composed of serous and/or mucus secretory epithelial cells. Serous secretory cells have an eosinophilic, slightly granular cytoplasm, whereas mucus cells have abundant foamy, lightly staining cytoplasm. The parotid gland is entirely composed of serous acini, the sublingual primarily mucous acini, and the submaxillary a combination of serous and mucous acini.

General histologic anatomy of the tubular digestive tract

Distal, or aboral, to the oral cavity is the tubular digestive tract. The general histologic organization of the tubular digestive tract, from esophagus to rectum, is quite similar. These organs have a central luminal space through which digesta passes.

The wall of these tubular organs are organized into distinct anatomic layers; from innermost (luminal) to outermost they are:

- Mucosa
- Submucosa
- Tunica muscularis, and
- Adventitia (serosa)

The mucosa includes both the lining epithelium as well as the lamina propria and muscularis mucosa. The collagen-rich lamina propria lies directly below the mucosal epithelium (the lamina propria is similar to the dermis of skin). The mucosa also contains a thin layer of smooth muscle, the muscularis mucosa, that separates the mucosa from the submucosa. The submucosa is composed of dense collagenous stroma and contains large blood vessels, lymphatics, and nerves. The tunica muscularis is composed of two distinct layers of smooth muscle: an inner circular layer and an outer longitudinal layer. Finally, the serosa is a thin layer of fibrovascular tissue that is, externally, lined by a thin layer of flattened, specialized epithelium called mesothelial cells.

The autonomic innervation of the gastrointestinal tract functions through two distinct regions containing autonomic ganglia: the submucosal (Meissener's) plexus, and the myenteric (Auerbach's) plexus. This autonomic innervation enables the waves of contraction that propel food through the tubular digestive tract, also known as peristalsis. These plexi are identified as discrete clusters of neurons and nerve fibers. The submucosal plexus is located in the submucosa, whereas the myenteric plexus is located between the inner circular and outer longitudinal smooth muscle layers of the tunica muscularis.

Esophagus

The primary physiologic role of the esophagus is the physical transport of ingesta from the oral cavity to the stomach. The esophageal mucosa is composed of a superficial layer of stratified squamous epithelium. Beneath the mucosa lie variable numbers of mucous glands, which produce mucus to lubricate and facilitate passage of digesta.

The constituents of the tunica muscularis varies based on species and location along the length of the esophagus. For example, in species such as the cat and horse, the tunic muscularis is primarily composed of skeletal muscle with distal segments incorporating smooth muscle. Dogs and ruminants have uniform skeletal muscle along the entire length. Pigs, similar to humans, primarily contain skeletal muscle in the orad 1/3 of the esophagus which transitions to a combination of skeletal and smooth muscle, and terminates in uniform smooth muscle, distally.

Chickens

Chickens have a specialized outpouching, or diverticulum, located in the distal esophagus (at the ventral base of the neck): the crop. The anatomy of the crop is identical to the esophagus. The crop serves as storage for ingesta, allowing for the chicken to consume a large amount of food and digest later.

Stomach

There is marked interspecies variations in the gastric compartment, exemplified by the difference between monogastrics (e.g. carnivores) and the multi-compartmentalized stomach of ruminants (e.g. cattle). The reader is referred to veterinary gross anatomy texts for the specifics pertaining to these structures, but the microscopic specifics will be noted here. Regardless, the primary function of the stomach is mechanical and chemical breakdown of digesta in preparation for absorption in the intestine.

Dogs and Cats

Domesticated carnivores (dogs and cats) have a single-compartment (simple) stomach that is entirely lined by glandular mucosa. The stomach is divided into several segments. From oral to aboral they are: cardia, fundus and pylorus.

The glandular stomach is so named due to the high secretory function of the mucosal epithelium, which produces mucus, hydrochloric acid, and pepsin. The mucosal surface is formed by numerous vertically oriented tubular glands. Superficially, the mucosa is composed of gastric pits, which are small mucosal depressions that form communications between the tubular glands of the mucosa and the gastric lumen. The gastric pits are primarily lined by mucus-secreting epithelial cells, columnar epithelial cells with pale to non-staining cytoplasm (mucus) and basal nuclei. These cells produce the mucus layer that coats the epithelial surface, protecting it from mechanical and chemical injury.

Below the gastric pits and comprising the majority of the mucosal epithelium within the fundus are two specialized epithelial cell populations. Parietal cells (oxyntic cells) are polygonal epithelial cells with abundant pale eosinophilic cytoplasm and round, central nuclei located within the central regions of the fundic mucosa. Parietal cells secrete the hydrochloric acid that acts as the primary chemical digestive effector of the stomach. Chief cells are low-columnar epithelial cells with granular eosinophilic cytoplasm and round, eccentric nuclei. Chief cells secrete pepsinogen, which is rapidly converted to the active form, pepsin, by hydrochloric acid. Pepsin is a proteolytic digestive enzyme (protease), thereby contributing to digestion. The distribution of the glandular epithelial constituents varies by region of the glandular stomach. The fundus contains large numbers of parietal cells, whereas the pylorus is rich in mucus cells but contains few parietal cells.

A third secretory epithelial cell type, gastrin cells, are located primarily within the pylorus. Gastrin cells secrete gastrin, a hormone induced by the presence of digesta within the stomach (e.g. eating) that stimulates secretion of the digestive enzyme pepsin (i.e. pepsinogen) and hydrochloric acid by chief and parietal cells, respectively. Gastrin cells are not readily discernible by routine histologic methods.

Similar to the other regions of the tubular digestive tract, a lamina propria underlies the gastric mucosa. External to the lamina propria are the gastric submucosa, tunica muscularis and serosa. The mucosa of the stomach may contain organized lymphoid aggregates, which are included in the gastrointestinal-associated lymphoid tissue (GALT; discussed later).

Stomach - ruminants (cattle, goats, sheep, cervids)

The ruminant stomach is divided into the nonglandular forestomach (rumen, reticulum, omasum) and the terminal glandular stomach, the abomasum. The distinct compartments of the ruminant forestomach warrant specific discussion here.

Rumen

The rumen is the largest (by volume) compartment of the ruminant forestomach. The primary function of the rumen is as a storage compartment to facilitate microbial (bacterial and protozoal) fermentation of ingesta. Volatile fatty acids are biproducts of this fermentation process and are readily absorbed across the ruminal mucosa into the circulation, where they serve as a directly accessible energy source. In addition, microbial digestion of fibrous ingesta begins the digestion necessary to access nutrients from ingesta that will undergo further breakdown in more distal parts of the gastrointestinal tract (e.g. abomasum).

The mucosa of the rumen varies regionally, but forms variably pronounced leaf-like papillae. The nonglandular mucosa of the rumen is composed of keratinizing stratified squamous epithelium. The rumen does not contain a muscularis mucosa.

Reticulum

The reticulum and omasum largely act to further mechanical dissociation of digesta following microbial breakdown. The mucosal surface of the reticulum is composed of long primary and shorter secondary folds that form a grossly apparent honeycomb-like pattern. As in the rumen, the reticular mucosa is composed of a keratinizing stratified squamous epithelium. The primary folds (large papillae) of the retriculum contain bundles of smooth muscle within the lamina propria, also referred to as the muscularis mucosa.

Omasum

The mucosal surface of the omasum is composed of large, thin, leaf-like structures called laminae. The microscopic structure of the omasum is similar to the reticulum in that it too contains a muscularis mucosa that extends into the laminae. In addition, the muscularis externa of the omasum extends into the laminae, which is unique to this compartment.

Abomasum

The terminal chamber of the stomach of ruminants, the abomasum, is functionally and microscopically analogous (identical) to that of the simple stomach of carnivores.

Stomach - additional interspecies variations

Horses

The equine stomach is composed of a proximal nonglandular mucosal region and an aboral glandular gastric mucosa. The anatomically distinct demarcation of these regions is referred to as the margo plicatus. The nonglandular mucosa is lined by stratified squamous epithelium, whereas the glandular portion is analogous to that of carnivores.

Pigs

In pigs, the gastric region immediately aboral to the esophageal sphincter is a well-delineated segment of nonglandular mucosa, the pars esophagea. The pars esophagea is lined by stratified squamous epithelium. The remaining gastric mucosa is glandular and analogous to that of carnivores.

South American Camelids (Llama, Alpaca)

Similar to the stomach of ruminants, the camelid stomach is anatomically divided into specialized compartments that serve to facilitate digestion of fibrous ingesta (foregut fermentation). The stomach of camelids is divided into three compartments: compartment 1 (C1), compartment 2 (C2), and compartment 3 (C3). Compartment 1 (C1) is analogous to the ruminant rumen, and is a large volume saccule that acts as a vat for microbial fermentation. In contrast to the rumen, C1 is lined primarily by glandular mucosa with regions of non-glandular stratified squamous epithelium. The C1 also contains numerous small out-pouchings, or saccules. The C2 is composed of both glandular mucosa and non-glandular stratified squamous epithelium. Finally, C3 is functionally and histologically analogous to the ruminant abomasum and canine stomach and is entirely glandular.

Chickens (and other fowl)

The proventriculus is glandular stomach that is functionally, but not histologically, analogous to the canine stomach. The wall of the proventriculus contains numerous large, well-circumscribed glands (proventricular glands), each formed by collections of branching tubules lined by regions of columnar mucous cells (microscopically similar to those of the gastric pits) and cuboidal to low columnar secretory cells. These secretory cells have eosinophilic granular cytoplasm and produce pepsinogen and hydrochloric acid, thereby serving the roles of both the chief and parietal cells of the canine stomach, respectively. The proventricular glands open to the luminal surface through apertures in the mucosal surface. The superficial proventricular mucosal epithelium is primarily composed of mucous cells.

The gizzard is the compartment immediately aboral to the proventriculus and is distinguished by a remarkably well-developed and thick smooth muscle wall that contributes to mechanical breakdown of ingesta. The gizzard is generally more developed in bird species with diets consisting of tough material (grains) and less so in species with more soft diets (fruit). Histologically, the mucosa is composed of vertically oriented tubular glands lined by columnar epithelium. The superficial (luminal) surface of the mucosa is lined by a thick layer of homogeneous, glassy eosinophilic to slightly basophilic material called koilin. Koilin forms a tough protective coating over the mucosal surface and contributes to the ability of the gizzard to grind particles of ingesta. The molecular composition of koilin is similar to but distinct from that of keratin. The koilin is produced by the tubular glands of the gizzard, which interdigitate with the koilin to provide structural integrity.

Small intestine

The small intestine of most domestic species is quite similar in function and histology. Structural and functional differences in specific regions of the small intestine impart differential functional capabilities to these segments. The small intestine is divided into three distinct segments, from oral to aboral: duodenum, jejunum, and ileum. The major functions of the small intestine are digestion, secretion, and absorption. The small intestinal mucosa has several anatomic adaptations that serve to create an immense surface area with which to digest and absorb nutrients. These include the plicae circulares (intestinal folds), villi, and microvilli.

The small intestinal mucosa is characterized by numerous, regularly distributed luminal papillary projections called villi. Villi are lined by columnar epithelial cells, enterocytes. Enterocytes have apical microvilli (brush border). Although individual microvilli are indistinguishable, the brush border is apparent as a faintly staining, uniform ~1 µm layer along the superficial surface of enterocytes. Enterocytes function mainly in digestion and absorption. The enterocytes are interspersed with goblet cells, columnar epithelial cells with abundant, poorly staining cytoplasm, representing mucin. The density of goblet cells is generally more abundant in more distal segments of the intestine.

The intestinal villi are contiguous with glands located at the base of villi: the crypts of Lieberkuhn, or intestinal crypts. The crypts contain the mitotically active population of intestinal epithelial stem cells. Mitotic figures are commonly seen in cells of the crypts. Within the crypts, epithelial stem cells divide and push upward (luminally), further differentiating into enterocytes or goblet cells. In this way, the small intestinal villi are similar to a production conveyor belt –intestinal epithelial cells are generated in the crypts and progressively migrate superficially along the villi towards the luminal surface where, at the tips of villi, the epithelial cells are sloughed into the lumen. This process occurs continually and promotes a high rate of enterocyte replacement/turnover.

In addition to enterocytes and goblet cells, the small intestine contains low numbers of accessory cells. In some species such as the horse, the crypts contain low numbers of cells, Paneth cells, that contain abundant eosinophilic cytoplasmic granules. These granules contain antimicrobial molecules important in gut innate immunity. Finally, low numbers of enteroendocrine cells are interspersed within the crypts. These enteroendocrine cells produce hormones that may include somatostatin, cholecystokinin, and secretin. Enteroendocrine cells are not readily apparent in routine H&E sections.

The lamina propria of the small intestine extends into and forms the core of small intestinal villi. The villous lamina propria is composed primarily of loose collagenous tissue, but contains a number of important structures and cells. The villous lamina propria is rich in both capillaries and lymphatics that help transport nutrients absorbed by enterocytes across the luminal surface. The small intestinal lamina propria also contains low numbers of immune cells, including lymphocytes and plasma cells, and small numbers of lymphocytes are regularly located within the villous epithelium.

Microscopic anatomic features

Although much of the previously discussed features of the small intestine apply to the duodenum, jejunum, and ileum, distinguishing microscopic anatomic features of the duodenum and ileum are detailed below.

Duodenum

The gastric pylorus empties into the lumen of the duodenum. The duodenal submucosa contains extensive tubuloacinar glands, Brunner's glands, that are lined by tall columnar epithelial cells with mucin-rich, poorly-staining cytoplasm. The Brunner's glands communicate with the lumen of the crypts of Lieberkuhn. The secretions of the Brunner's glands are alkaline and help to neutralize the acidic digesta received from the stomach.

The pancreatic duct and common bile duct insert into the wall of the duodenum and communicate with the duodenal lumen.

Ileum

The ileal mucosa contains large numbers of organized lymphoid tissue (lymphoid follicles), termed Peyer's patches. The Peyer's patches serve as both a primary and secondary lymphoid organ. Peyer's patches are critical components of the GALT.

Cecum and large Intestine

Similar to the stomach, the cecum and large intestine have a wide variation in size, shape and function in domestic species. In species such as the cat, the cecum is small and relatively inconspicuous, whereas species such as the horse and rabbit have large, well-developed ceca. This is largely attributed to function: the horse and rabbit are hindgut fermenters. Regardless of the functional and gross anatomic variations, the mucosas of the cecum and large intestine of most species are histologically similar.

Unlike the small intestine, the cecal and large intestinal mucosa lack villi. Instead, the mucosa is composed of densely arranged straight tubular glands (colonic glands). Colonic glands are lined by enterocytes and goblet cells (see small intestine) with a much higher density of goblet cells than in the small intestine. The base (crypts) of the glands are analogous to the small intestinal crypts of Lieberkuhn and epithelial proliferation occurs in this population of epithelial cells.

Similar to the ileal Peyer's patches, the cecum and large intestine contains abundant, well-organized lymphoid tissue in the mucosa.

The rectal mucosa is similar to the large intestine in histologic appearance.

Chapter-9

Hepatobiliary System

The liver is the largest visceral organ in the body. This chapter will cover the hepatobiliary system.

Chapter Learning Objectives

By the end of this chapter, you should be able to do the following:

- ❖ Describe the role of the liver in the synthesis of lipids and proteins.
- ❖ Understand the role and importance of hepatocytes and in xenobiotic bio-transformation.
- ❖ Compare and contrast the location, function, and structural (microscopic) features of
- ❖ hepatocytes, Kupffer cells, Ito cells and Pit (immune) cells.
- ❖ Describe the hepatic blood supply and diagram the flow of blood in the liver.
- ❖ Describe and identify the three major structures that comprise the portal triad: portal vein, hepatic artery, and bile ductule.
- ❖ Describe the flow of bile and the macro- and microscopic structures through which it flows,
- ❖ from hepatocyte to gastrointestinal tract.
- ❖ Illustrate the different schemes used to describe the organization structure of the liver.

Review Questions

By the end of this chapter, you should be able to answer the following:

- ❖ What regions of the liver are most susceptible to hypoxemia (low blood oxygen)
- ❖ and why?
- ❖ A porto-systemic shunt is formed when the portal circulation bypasses the liver and, instead, flows into the systemic venous circulation (most frequently as the result of a congenital vascular malformation). What are some possible consequences of blood from the portal circulation bypassing the liver? (Hint: recall the normal physiologic functions of the liver) (additional reading available through OSU libraries)

Overview

The liver is the largest visceral organ in the body, accounting for roughly 1-4% of total body weight, depending on species (3-4% in carnviores and 1-2% in herbivores). The liver is arranged into distinct lobes, each of which is encased by a thin, translucent layer of mesothelium (Glisson's capsule). The number of liver lobes varies among different species. Canines and felines have five liver lobes, and herbivores have four liver lobes. The number of liver lobes does not represent function but is a reflection of species diversification.

Hepatic functions

Detailed discussion of the hepatic physiology is beyond the scope of this book. However, understanding the roles of the liver is necessary when considering the microscopic anatomy.

Nutritional metabolic functions

The liver functions in the modification and transport of nutritional molecules, and therefore plays a critical role in energy metabolism. This includes the generation, breakdown, modification, and storage of nutritional molecules including glucose/glycogen, proteins, and lipids.

Xenobiotic metabolism

Xenobiotics include any compounds that are "foreign" to an animal. An example of xenobiotics are medications such as the class pain relief drugs referred to as NSAIDs (NSAIDs = non-steroidal anti- inflammatory drugs), but xenobiotics also include non-therapeutic compounds.

Xenobiotics are metabolized by hepatocytes via biotransformation, the process by which xenobiotics are chemically modified and excreted into bile. Hepatocytes are rich in cytochrome P450, a metabolic enzyme responsible for biotransformation. Biotransformation of a xenobiotic aims to render the agent less toxic and soluble for excretion. Occasionally, activation of the xenobiotic during biotransformation may result in an intermediate compound that is *more* toxic than the original form, leading to toxicity! Nonetheless, hepatocytes are the workhorses of xenobiotic biotransformation, and as such, they serve a major protective role in preventing the build-up and toxicity of xenobiotics.

Production of circulating plasma proteins

Hepatocytes produce many of the most abundant circulating plasma proteins including albumin, lipoproteins (LDL, VLDL, and HDL), and glycoproteins. These hepatocyte-produced proteins function in diverse areas such as energy and molecular transport and inflammation. The liver is also the primary source of proteins of the coagulation cascade. Hepatocytes produce almost all of the clotting factors (except Factor VIII, produced by endothelium) required in the coagulation cascade, including fibrinogen, prothrombin, antithrombin III, Protein C and Protein S.

Storage and conversion of vitamins and minerals

The liver is a major site of storage and conversion of various vitamins and minerals that include Vitamin A (retinol), Vitamin D, Vitamin B12, Vitamin K, iron and copper.

Bilirubin metabolism

The liver receives red blood cell fragments from the spleen and further processes these products for recycling or excretion. Hemoglobin from red blood cells is broken down into heme and globin. The iron containing heme is further processed into bilirubin. Hepatocytes conjugate bilirubin, converting uncojugated (insoluble) bilirubin to conjugated (soluble) bilirubin, and is excreted and incorporated into bile to be excreted into the gastrointestinal tract.

Cells of the liver

Hepatocytes

The liver parenchyma is primarily comprised of hepatocytes. Hepatocytes are polygonal epithelial cells with abundant eosinophilic, granular cytoplasm and large, centrally located round nuclei. Hepatocyte nuclei often contain a prominent nucleolus. Binucleated hepatocytes (= containing two nuclei) are not uncommon.

Hepatocytes contain many mitochondria and have extensive smooth and rough endoplasmic reticulum and Golgi apparatus, all of which contribute to the eosinophilic staining of their cytoplasm. However, the microscopic appearance ofhepatocyte cytoplasm can vary based on the nutritional status of the animal (recall that the liver plays a major role in energy and nutrient metabolism). The most common cytoplasmic changes are due to intracytoplasmic glycogen and lipid. Histologically, glycogen appears as irregular, poorly-defined non-staining (clear) spaces within the cytoplasm. In contrast, lipid appears as either a single, large non-staining, sharply-defined cytoplasmic vacuole or as multiple variably-sized non-staining vacuoles. Both glycogen and lipid are largely "washed-out" during routine processing of histology slides, which accounts for their lack of staining in tissue section. While both of these changes can be physiologically normal (or clinically inconsequential), excessive accumulation of lipid within hepatocytes can be pathologic, and is referred to as hepatic lipidosis.

Hepatocytes are arranged in radiating cords (hepatic cords) approximately one to two cells in thickness. Adjacent hepatic cords are separated by vessels: the sinusoidal capillaries (sinusoids). These capillaries are lined by endothelial cells, which are difficult to appreciate in normal histologic section due to their flattened nuclei. More on hepatic architecture as well as the vasculature and blood supply to the liver will be discussed later in this chapter.

Ito cells

Ito cells are also known as stellate cells, fat storing cells, or lipocytes. Ito cells reside in the perisinusoidal region known as the space of Disse. The space of Disse is the narrow region located between endothelial cells and hepatocytes. These cells are identified histologically by their large lipid vacuoles. Ito cells function in the uptake, storage and maintenance of vitamin A (retinol), as well as the production of extracellular matrix (collagen types I and III), regulation of sinusoidal blood flow, and hepatic tissue repair following injury.

Kupffer cells

Kupffer cells are phagocytes ("phago" = eating, "cyto" = cells) derived from monocytes and located within the vascular spaces of hepatic sinusoids lining the endothelial surfaces. These Kupffer cells function in removing aged red blood cells from circulation and in phagocytizing and removing blood- borne microbes or endotoxins absorbed from the gastrointestinal tract. Histologically, Kupffer cells are individualized round cells within sinusoids that are difficult to identify on histology unless they contain phagocytized cytoplasmic material, such as red blood cells.

Oval cells

Oval cells are pluripotent stem cells that, due to their ability to differentiate into several different cell types ("pluri" = several), serve a primary role in the repopulation of hepatocytes and other hepatic cells (e.g. biliary epithelium) following hepatic injury. These cells are not readily identifiable in normal tissue sections.

Pit cells

Pit cells are short-lived granular lymphocytes that reside within hepatic sinusoids and contribute to immunity.

Hepatic blood supply

The liver is unique in that it receives blood from two sources: the hepatic artery and the portal vein. As these vessels enter the liver, their terminal branches run alongside branches of the bile ducts and course together throughout the liver parenchyma within portal triads (triad = three = hepatic artery, portal vein, bile ductule). **Note that lymphatics also run within portal triads.**

Portal veins are identified as larger, blood-filled spaces with a thin to inapparent vessel wall. Hepatic arteries are smaller in diameter but have a more well-developed smooth muscle wall. Blood from both the hepatic arteries and portal veins empty into the adjacent peri-portal hepatic sinusoids and flows towards the central vein.

The portal vein supplies the majority of blood to the liver. The portal circulation is a venous system that drains blood from the gastrointestinal tract, gall bladder, spleen, and pancreas. Venous blood coming from the gastrointestinal tract contains not only digested nutrients, but also toxins and, frequently, microbes. A major role of the liver in receiving this portal circulation is to metabolize these nutrients and eliminate or mitigate these toxins and microbes.

The hepatic artery delivers highly-oxygenated blood to the liver. However, once within the liver, blood from the hepatic artery and the portal vein empty into the hepatic sinusoids, combining highly oxygenated blood (hepatic arterial supply) with poorly oxygenated blood (portal vein supply). As a result, hepatocytes receive blood with reduced oxygenation. Further, when considering that blood flows from the portal triad towards the central vein, blood arriving to hepatocytes surrounding the central vein is relatively poorly oxygenated. These concepts are important in the understanding of liver pathology and how regions of the liver are differentially more susceptible to hypoxia.

The hepatic sinusoid endothelium is fenestrated, containing spaces between adjacent endothelial cells. These gaps in the endothelium allow passage of nutrients and other small compounds into the perisinusoidal space (space of Disse) where hepatocytes are able to take up these compounds. Microvilli on the surface of hepatocytes provide additional surface area for absorption of these molecules. Additionally, proteins and lipoproteins produced by hepatocytes are transferred into circulation in these perisinusoidal spaces.

The biliary system

Production and transport of bile

The main function of the biliary system is the transportation of bile from the liver to the gut. Bile, produced by hepatocytes, is composed of water, cholesterol, bile pigments (bilirubin), electrolytes (sodium, potassium, chloride, bicarbonate, etc.), and bile salts. Bile salts (bile acids) serve as emulsifying agents useful for the digestion and absorption of lipids and lipid soluble vitamins (e.g. vimatin A and vitamin K) from the intestinal tract.

In contrast to blood flow, the direction of bile flow is from centrilobular (nearer the central vein) towards the portal triads. Bile is produced by hepatocytes and secreted into bile canaliculi, narrow channels formed between the cellular membranes of adjacent hepatocytes. These bile canaliculi are extensive and connect to form a highway for bile transport, all flowing towards the portal triads. As the bile canaliculi join near the periphery of the lobules, they form into larger canals (also known as the Canals of Hering). The Canals of Hering empty into bile ductules, located within portal triads. Each portal triad contains one or several bile ductules.

Bile ductules have a central lumen lined by a single layer of cuboidal epithelial cells. Bile ductules from portal triads course through the liver and connect to form increasingly larger caliber structures, bile ducts, that are lined by short to tall columnar epithelium. These bile ducts will eventually empty into the hepatic duct, which empties into the cystic duct and gallbladder (dogs, cats, ruminants) or common bile duct (species that lack gallbladders, such as horses and alpacas).

Gallbladder

The gallbladder is a blind-ended, sac-like organ attached to the visceral surface of the liver. The gallbladder receives, stores and concentrates bile entering from the liver. In response to a fatty meal, the enteroendocrine cells of the duodenum secrete cholecystokinin, a hormone that stimulates smooth muscle contraction and release of bile from the gall bladder.

The structure of the gallbladder wall is multilayered, similar to the intestinal tract. The luminal mucosal surface is lined by simple columnar epithelium. Beneath the epithelium lies a vascular-rich lamina propria. External to the lamina propria is the muscularis externa, composed of randomly oriented smooth muscle bundles. Occasionally, diverticula or invaginations of the mucosa can extend in the muscularis externa. External to the muscularis externa lies the adventitia.

Structural organization of the liver

Knowing the individual cellular and structural components of the liver, we now consider the structural organization of the liver. There are two useful ways to describe the histological architecture of the liver: lobules or acini.

Hepatic lobule (Kiernan's lobule)

The lobular organization of the liver is one of the more histologically relevant ways to understand hepatic architecture. The two major structural landmarks to consider when reviewing the organizational aspects described below are the central veins and the portal triads. The central vein lies within the central of the hepatic lobule. Each central vein has multiple portal triads that provide blood supply to that region. Therefore, you can easily orient yourself by first identifying either a central vein (the center of a lobule) or identifying a portal tract (the periphery of a lobule).

The hepatic lobule is observed as a hexagon-like structure where the central vein is located at the center and the hepatocytes form radiating cords from the central vein, similar to the spokes of a wheel radiating from the central hub. At the outside "corners" of the hexagon are portal triads.

The hepatocellular regions within the classical lobule include *portal, midzonal* and *centrilobular*. For instance, portal hepatocytes are those closest to the portal triads, and centrilobular hepatocytes surround the central vein.

The pig liver contains increased amounts of fibrous connective tissue that bridges portal tracts, forming a band that highlights the perimeter of the lobule and illustrates the lobular organization of the liver.

Hepatic acinus (Acinus of Rappaport)

Whereas the lobular organization is based on the physical organization of structures within the liver, the hepatic acinus is based more on function. The acinus (plural: acini) highlights the hepatic blood flow and metabolic activity, and is particularly useful when considered liver disease or pathology.

The acinus is shaped as a triangle. The central vein is located at the point of the triangle and the base is at the outer edge, with the other two point of the triangle associated with portal triads. In this model, hepatocytes are defined by zones. Hepatocytes in zone 1 (also called the periportal zone) are located closest to the portal areas and are the first to receive oxygen and nutrient rich blood. These hepatocytes are metabolically active and involved in cholesterol synthesis, fatty acid oxidation and bile acid production. In contrast, hepatocytes in zone 3 (centrilobular) receive oxygen and nutrient depleted blood and are a major effector of glycolysis, lipogenesis, and xenobiotic biotransformation (highest cytochrome p450 concentration). Zone 2 (midzonal) hepatocytes have considerable regenerative potential and this region contains the greatest number of oval cells.

Chapter-10

Respiratory System

The primary functions of the respiratory system are 1) smell, 2) air conduction, and 3) the exchange of oxygen and carbon dioxide between the animal and the environment, referred to as respiration. This chapter will cover the respiratory system.

Chapter Learning Objectives

By the end of this chapter, you should be able to do the following:

- ❖ Compare and contrast the histologic features, cell composition and functional differences between respiratory epithelium and olfactory epithelium.
- ❖ Describe the histologic features and functions of the vomeronasal organ.
- ❖ List four functions of the conducting airways.
- ❖ Compare and contrast the histologic features of the trachea, bronchi, and bronchioles.
- ❖ Describe the structures that form the alveolar air-blood barrier.
- ❖ Compare and contrast the histologic features and functions of the alveolar cells: type I pneumocytes, type II pneumocytes, and alveolar macrophages.

Review Questions

By the end of this chapter, you should be able to answer the following:

- ❖ Respiratory epithelium is common to the upper respiratory tract (e.g. nasal cavity) and lower respiratory tract (e.g. bronchi). What are the primary functions of respiratory epithelium?
- ❖ What histologic features distinguish the trachea from the mainstem bronchi? What features distinguish mainstem bronchi from bronchioles?
- ❖ What are two separate disease processes that could contribute to the loss of smell (Hint: consider the "peripheral" aspects of the sense of smell as well as the "central" aspects).
- ❖ What are the normal functions of the mucociliary escalator and which cells contribute to this function? What would be the consequences of damage to or reduction in the efficiency of the mucociliary escalator?
- ❖ What structural features of alveoli facilitate gas exchange?

Overview

The primary functions of the respiratory system are 1) smell, 2) air conduction, and 3) the exchange of oxygen and carbon dioxide between the animal and the environment, referred to as respiration. Respiration occurs at the air-blood interface, which lies at the most distal part of the respiratory tract. Although this exchange of oxygen and carbon dioxide is the ultimate functional goal of the respiratory tract, air must first be transported to this air-blood interface. Therefore, the two crucial functions of the respiratory tract are air conduction and respiration. The importance of considering these two roles is apparent during respiratory diseases in which interference with either air conduction (such as obstructive respiratory disease) or the air-blood interface (such as bacterial bronchopneumonia) result in life-threatening hypoxia.

Clinical discussion of the respiratory tract is generally divided into the upper respiratory tract and lower respiratory tract. The upper respiratory tract is comprised of the nasal cavity and nasopharynx, whereas the lower respiratory tract includes the larynx, trachea, bronchi, bronchioles, and alveoli. Histologically, there is much overlap between the upper and lower respiratory tract, and therefore, the distinction of upper vs lower respiratory tract is not reinforced in the discussion to follow. Instead, regions of air conduction and gas exchange have important functional and histologic distinctions, and this categorization will be utilized for the purpose of the discussion to follow.

Air Conduction: Nasal Cavity, Paranasal Sinuses, And Vomeronasal Organ

Nasal Cavity

The paired external apertures of the nasal cavities in our domestic species are termed the external nares (nostrils). During inspiration, air enters through the nares into the most rostral segment of the nasal cavity: the nasal vestibule. The vestibule of most species is lined by keratinized stratified squamous epithelium that may contain hair follicles and other adnexa.

The stratified squamous epithelium of the vestibule abruptly transitions into to the characteristic respiratory epithelium. Traditional respiratory epithelium is ciliated, pseudostratified columnar epithelium. Respiratory epithelium is primarily composed of these ciliated columnar epithelial cells with basal nuclei and interspersed with moderate numbers of goblet cells. Goblet cells secrete mucus, and are characterized by cytoplasm filled with poorly staining, basophilic material (mucin).

Respiratory epithelium lines the luminal surface of the nasal cavity, including the nasal turbinates. The nasal turbinates are scroll-like structures within the nasal cavity that are formed by a thin core of bone. The lamina propria and submucosa lie below the respiratory epithelium. The lamina propria may contain variable numbers of tubuloalveolar secretory glands. Similar to the salivary glands of the gastrointestinal tract, these glands may be comprised of serous cells, mucous cells, or a mixture of serous and mucous cells. These glands secrete into ducts that communicate with the surface epithelium.

Although the majority of the surface area of the nasal cavity mucosa is respiratory epithelium, portions of the more caudal regions of the nasal cavity are covered by olfactory epithelium. Whereas most of the nasal cavity is directed at physical air conduction, olfactory epithelium is a specialized epithelium that facilitates the sense of smell. Olfactory epithelium is tall, pseudostratified epithelium that contains three distinct cell types: olfactory receptor cells, sustentacular cells, and basal cells.

Olfactory receptor cells are modified neurons and chemoreceptor cells interspersed within the olfactory epithelium. The cell bodies of these olfactory receptor cells reside within the epithelium and have apical cilia that project onto the surface of the epithelium. These cilia are not readily distinguishable on microscopy. Nerve fibers (axons) extend from the olfactory receptor cells into the lamina propria and synapse with other axons to form collectively form olfactory nerves. These olfactory nerves, in turn, pass through perforations in the cribriform plate of the skull and enter the olfactory bulb of the brain. In this way, odors are detected by olfactory receptor cells and transmitted via axons to the brain where these odors are converted to the sense of smell.

Sustentacular cells are histologically similar to respiratory epithelium and have apical microvilli.

These cells provide structural and metabolic support to the epithelium. Basal cells are cuboidal to short polygonal cells that reside adjacent to basement membrane.

Paranasal sinuses

The sinuses are bone-encased, air-filled spaces within the skull that communicate with the nasal cavity. Species such as the horse have well-developed and distinct paranasal sinuses that include the frontal, dorsal conchal, ventral conchal, rostral maxillary, caudal maxillary, and sphenopalatine. Species such as the dog have only frontal and maxillary sinuses. Regardless, the paranasal sinuses are lined by respiratory epithelium similar to that elsewhere in the nasal cavity and respiratory tract.

Vomeronasal organ

The vomeronasal organs are paired (left and right) specialized chemoreceptor organs adjacent to the nasal septum within the nasal cavity of domestic species. Structurally, the vomeronasal organ is a thick layer of sensory epithelium partially encased by cartilage and associated with an abundant vascular supply. Histologically, the sensory epithelium is similar to that of olfactory epithelium, with a thick pseudostratified epithelium that contains three populations of cells: receptor cells, sustentacular cells, and basal cells. The vomeronasal receptor cells, similar to olfactory receptor cells, are modified neurons that detect pheromones (e.g. pheromones in urine) and transmit to the brain via efferent axons. Access of air to this region of the nasal cavity is facilitated through passage through ducts in the dorsal aspect of the oral cavity (roof) of some species by the behavioral action of the flehmen response. This behavior looks different in different species. Examples include the curling of the upper lip in ungulates, chattering of the mouth in dogs (tonguing), and the open-mouthed posture of cats.

Air conduction: larynx, trachea, bronchi and bronchioles

Larynx

The larynx is a complex, tubular organ, the structure of which is primarily composed of elastic and hyaline cartilage and skeletal muscle. The larynx connects the naso- and oro-pharynx with the trachea, functioning in air conduction, vocalization, and in obstructing passage of ingesta into the trachea during deglutition. The epithelium of the larynx is variable, but is generally stratified squamous epithelium in the rostral segment proximal to the pharynx, and ciliated pseudostratified columnar (respiratory) epithelium elsewhere. The lamina propria subjacent to the laryngeal epithelium contains secretory glands.

Trachea, bronchi and bronchioles

The luminal (mucosal) surface of the trachea is lined by ciliated, pseudostratified columnar epithelium (i.e. respiratory epithelium) interspersed with goblet cells. The lamina propria and submucosa also contain small mixed (mucous and serous) glands and smooth muscle. The trachea is supported by a series of C-shaped cartilaginous (hyaline cartilage) rings. Dorsally, the ends of the incomplete cartilaginous rings of the trachea are connected by a band of smooth muscle: the trachealis muscle. Externally, the trachea is surrounded by a thin layer of fibrovascular connective tissue (adventitia).

Distally, the trachea divides into the left and right mainstem bronchi. Bronchi are histologically similar to the trachea. They are lined by ciliated, pseudostratified columnar epithelium (respiratory epithelium) and interspersed with goblet cells. The walls of bronchi are also supported by cartilage and smooth muscle. The smooth muscle of airways mediates bronchoconstriction, the process of contraction and subsequent narrowing of the airway lumen. Bronchoconstriction is mediated, in part, via mast cell-derive histamine, and bronchoconstriction is a life-threatening effect of massive histamine release (e.g. anaphylaxis). Finally, the lamina propria of bronchi also contain moderate numbers of mixed glands. These are particularly abundant in cats. As the bronchi extend deep into the pulmonary parenchyma, they continue to bifurcate, giving rise to smaller caliber (diameter) secondary and tertiary bronchi.

Eventually, bronchi give way to smaller conducting airways, bronchioles. Bronchioles are thin- walled airways lined by variably ciliated or non-ciliated columnar epithelium. The walls are composed of tangentially arranged smooth muscle. Bronchioles lack cartilage and glands which, along with their smaller size, distinguish them histologically from bronchi.

The transition between regions of air conduction (i.e. bronchioles) and regions of air exchange (i.e. alveoli) in the lung progresses through sequentially distinct segments; from proximal to distal: terminal bronchioles, respiratory bronchioles, alveolar ducts, alveolar sacs, and alveoli.

Air conduction: terminal bronchioles, respiratory bronchioles and alveolar ducts

Terminal bronchioles, respiratory bronchioles and alveolar ducts

Bronchioles open into short segments called terminal bronchioles, which are thin-walled branches of the bronchioles. Terminal bronchioles transition into respiratory bronchioles. Respiratory bronchioles are lined by two types of epithelial cells: ciliated columnar cells and club cells (also known as Clara cells). Histologically, club cells are cuboidal epithelial cells that lack cilia but contain microvilli (difficult to appreciate microvilli on histopathology). Club cells are rich in cytochrome P450, an enzyme responsible for the biotransformation of inhaled and circulating toxins. P450 is also abundant and highly active in hepatocytes, another cell type responsible for the biotransformation of toxins. Secondary functions of the club cells include production of pulmonary secretions and as progenitor (proliferating) cells that may repopulate after cell loss and/or injury.

The walls of respiratory bronchioles are discontinuous, forming openings into alveolar ducts. Alveolar ducts are conducting regions are lined by knob-like projections of cuboidal epithelium that cover small amounts of smooth muscle. These projections of cuboidal epithelium are often located on either side of the entrance into alveolar sacs, which branch from the "walls" of the alveolar ducts. Alveolar sacs are indistinct segments that open into individual alveoli.

Gas exchange: alveoli

Alveoli

Alveoli (singular: alveolus) are the site of gas exchange in the lung. Alveoli are thin-walled, sac- like structures lined by a single layer of flattened squamous epithelial cells: type I pneumocytes. This extremely thin nature of the type I pneumocytes facilitates gas exchange across their surface. The alveolar wall, or septum, consists of capillaries and minimal connective tissue support. As such, alveolar septal capillaries are almost in direct apposition with the type I pneumocyte. The type I pneumocyte and capillary lumen are separated only by the basement membrane of the type I pneumocytes, minimal or absent septal connective tissue, the basement membrane of the capillary endothelium, and the endothelial cells themselves. This provides an extremely narrow gap through which gases can diffuse, providing for efficient exchange of oxygen and carbon dioxide between alveolar spaces and capillaries.

In addition to the type I pneumocytes, alveoli also contain several additional cell types. Type II pneumocytes are cuboidal epithelial cells frequently residing in the corners of alveolar spaces. Type II pneumocytes are responsible for the secretion of surfactant. Surfactant is a fluid composed of phospholipids and proteins that coats the surface of alveolar spaces to reduce surface tension, allowing for alveoli to expand and remain open. In addition, type II pneumocytes play a crucial role in tissue repair in the lung. Following damage to type I pneumocytes, type II pneumocytes proliferate and differentiate into type I pneumocytes, thereby restoring alveolar structure.

Finally, alveolar spaces contain a resident population of macrophages, alveolar macrophages. Normally, these macrophages are located in close apposition to type I pneumocytes, and can be mistaken for type II pneumocytes. These macrophages, which are relatively low in number, readily phagocytize debris within alveoli, which can include surfactant, edema fluid, red blood cells, and pathogens (e.g. bacteria). Low numbers of macrophages are also normally present within the alveolar septal interstitium.

Pleura

Pulmonary pleura (Visceral pleura)

The outer surface of the entire lung parenchyma is covered by the visceral pleura. The visceral pleura is composed of a thin, loose connective tissue. The outer surface is lined by specialized squamous-like cells, mesothelium. Mesothelial cells appear histologically similar to epithelial but are mesoderm derived, and express proteins consistent with both mesenchymal (stromal) cells and epithelial cells. The mesothelium produces a lubricating substance that acts to reduce friction between the visceral pleura and the pleura coating the thoracic body wall (parietal pleura) during respiration. The connective tissue of the visceral pleura is contiguous with the connective tissue of the pulmonary lobular septae that course through the pulmonary parenchyma.

Pulmonary blood supply

Blood supply

The lungs receive two distinct sources of blood supply: pulmonary arterial supply and bronchial arterial supply. The pulmonary arterial circulation originates from the right side of the heart. Histologically, pulmonary arteries and veins are often adjacent to bronchi and larger bronchioles. These vessels progressively divide into smaller caliber vessels and, terminally, form a network of capillaries that course through alveolar septa and function in gas exchange. These capillaries then form into venules, then veins that progressively become larger in size and exit the lung, entering the left atrium. Bronchial arteries originate from the thoracic aorta and also course along large airways, providing much of the blood supply to these structures, as well as the pleura.

Chapter-11

Urinary system

This chapter will cover the urinary system.

Chapter Learning Objectives

By the end of this chapter, you should be able to do the following:

- ❖ Describe the components of the nephron, with specific reference to histologic features, location and function.
- ❖ Identify and describe the histologic features, location and function of the following components of the glomerulus: endothelial cells, glomerular basement membrane, podocytes, mesangial cells, Bowman's space, Bowman's capsule.
- ❖ Describe how plasma (intravascular) is filtered across the glomerular basement membrane, and describe the cells that contribute to the filtration barrier.
- ❖ Describe how the tubular segment of the nephron modifies filtrate to maintain homeostasis.
- ❖ Describe the anatomic and microanatomy of the lower urinary tract.

Review Questions

By the end of this chapter, you should be able to answer the following:

- ❖ Define the following renal components: Medullary ray, lobe, lobule, papilla and calyx
- ❖ How do the cortex, corticomedullary junction, and medulla differ?
- ❖ What are the 3 components of the glomerular filtration barrier? What are their properties?
- ❖ Describe endothelial fenestrations and glycocalyx.
- ❖ Describe podocytes (their shape) and slit diaphragms.
- ❖ Describe reabsorption. What types of molecules are reabsorbed? How are they reabsorbed?
- ❖ What are the distinguishing features of the proximal tubules and distal convoluted tubules?

What are some mechanisms for transporting molecules across a cell membrane or space? Which of these require energy, and why?

- ❖ What structures are lined by urothelium?

Overview and anatomy

Urine is formed in the kidney and delivered to the bladder by ureters. Urine exits the urinary bladder through the urethra. On longitudinal section, the kidney can be divided into the cortex, medulla, renal papilla and renal pelvis. The concave border of the kidney is known as the hilum (or hilus), where the renal vessels (artery and vein) and nerves enter the kidney. At this location, the renal pelvis funnels into the ureter.

In addition to dividing the kidney into cortex and medulla, it can also be separated into lobes and lobules as well as medullary rays and the cortical labyrinth. Lobes are easiest to visualize in cattle and pigs wherein every single renal papilla drains the overlying medulla and surrounding cortex. In unipapillary kidneys (e.g. carnivores), the distinction between lobes is more difficult to appreciate. A lobe can be further divided into lobules, which are composed of a single medullary ray and the surrounding cortical labyrinth, both of which are described in more detail later.

Normally, the cortex comprises approximately 2/3 of the renal parenchyma and contains most of the glomeruli and numerous cross sections of tubules. The medulla only has tubules and duct and does not contain glomeruli.

An ultrafiltrate of urine is formed in the cortex and is modified during its course through the medulla. The structures within the cortex, primarily the glomeruli, drive filtration and the tubules within the medulla are responsible for reabsorbing, excreting and concentrating the urinary filtrate.

The nephron is the functional anatomic unit of the kidney. Each nephron has a glomerulus and a long tubule which becomes a duct before emptying into the pelvis. The glomeruli, drive filtration whereas the tubules are responsible for reabsorbing, excreting and concentrating the urinary filtrate. The main components of the nephron include:

- ❖ Glomerulus
- ❖ Bowman's space (surrounded by Bowman's capsule)
- ❖ Proximal convoluted tubule
- ❖ Proximal straight tubule
- ❖ Loop of Henle (descending limb and ascending limb)
- ❖ Distal convoluted tubule
- ❖ Collecting duct (cortical and medullary)
- ❖ Papillary duct

The glomerulus

The glomerulus is a complex web of capillaries derived from the afferent arteriole. Glomeruli can be located in the cortex of the kidney or the corticomedullary junction.

Blood that is traveling through the capillary loops is selectively filtered across the glomerular filtration barrier to form an ultrafiltrate of the plasma. This filtration barrier permits passage of small and some medium sized molecules while blocking the passage of large molecules. Filtration is driven by hydrostatic and oncotic pressure. The former pushes fluid and solutes out of the capillary lumen, whereas the latter retains fluid. The major players of the glomerular filtration barrier include: endothelium, glomerular basement membrane, and podocytes.

The glomerular endothelial cells are characterized by numerous fenestrations. Using high magnification with transmission electron microscopy (TEM), one can visualize the endothelial cytoplasm, which has numerous fenestrations or transcytoplasmic holes that allow passage of the plasma from the vasculature to the interstitium. Another important component of the endothelial cell is the glycocalyx, which covers the luminal surface of the cell. The glycocalyx is composed of negatively charged glycoproteins and proteoglycans (sugar-protein molecules). filtering out negatively charged molecules, leukocytes, red blood cells, platelets and larger proteins.

The glomerular basement membrane (GBM) is a product of the fusion of the basement membrane produced by endothelial cells and podocytes (described below). Using TEM, one can see that the GBM is trilaminar (i.e., three-layered), and these components include, from the endothelial to the podocyte aspect, the lamina rara interna, lamina densa, and lamina rara externa. This part of the filtration barrier is also negatively charged and restricts passage of larger proteins.

Podocytes are epithelial cells. Aside from providing structural and functional stability of the glomerulus, the podocyte is a vital component of the glomerular filtration barrier. They are located on the "abluminal" (i.e., opposite of capillary lumen) surface of the capillary loop. Podocytes have numerous processes, similar to tentacles, that surround the capillaries. Each process has many smaller processes (foot processes or pedicles) that are oriented perpendicularly to the capillary wall and attach to the underlying GBM. The foot processes interdigitate with foot processes of adjacent podocytes. The space between the foot processes forms the slit diaphragm. The slit diaphragm is a molecular sieve, and it provides an additional filtration barrier based mainly on size exclusion. Aside from being part of the filtration barrier, podocytes also help maintain the shape of the capillary loop shape by counteracting intraglomerular pressure. Lastly, podocytes produce vascular endothelial growth factor (VEGF) to which is a molecule that the glomerular endothelial cell needs to survive.

Additional structures of the glomerulus

Mesangium

Collectively, the mesangial cells and the glomerular extracellular matrix that they produce constitute the mesangium. The mesangium provides structural integrity to the glomerular tuft and modulates glomerular perfusion through smooth muscle-like activity. These cells are also involved in the production and secretion of cytokines in response to glomerular injury. Histologically, there are 2 to 3 mesangial cell nuclei beneath each capillary loop.

Bowman's space and Bowman's capsule

Once the ultrafiltrate is formed, the fluid enters Bowman's space, the space between the capillary tuft and Bowman's capsule. Bowman's capsule is lined by parietal epithelium, which is simple squamous epithelium. The ultrafiltrate drains toward the urinary pole and enters the proximal convoluted tubule.

Efferent/ Afferent Arterioles

The blood enters the glomerular tuft via the afferent arteriole and exits via the efferent arteriole. These vessels are immediately adjacent to one another at the vascular pole, which is at the opposite end of the glomerulus from the urinary pole. These vessels can contract or dilate, changing the amount of blood that flows into the glomerular tuft. Importantly, the contraction can occur in response to molecular signals from the immediately adjacent tubules.

Renal tubules: mechanisms of reabsorption

The kidneys filter/excrete waste and reabsorb essential electrolytes, nutrients, and fluid (ultimately to maintain homeostasis). Tubular reabsorption occurs via multiple mechanisms of transport.

Mechanisms of tubular reabsorption

Passive transport

Passive transport is movement of a molecule without utilizing energy. In this form of transport, solutes move from a high concentration to low concentration (along their concentration gradient).

Active transport

Active transport requires energy in order to transfer a solute *against* an electrochemical gradient. There are two major forms of active transport in the tubules:

- *Primary active transport* involves movement of a solute against an electrochemical gradient at the expense of ATP (energy). An example of this type of transport is seen with the Na+/K+ ATPase pump on the basolateral cell membranes of the tubules. Sodium (low intracellular concentration, high extracellular concentration)is pumped out of the cell into the interstitium. At the same time, potassium is moved into the cell from the interstitium, also against its concentration gradient.
- *Secondary active transport* is a form of active transport where two different molecules interact with a single membrane transporter to be translocated across the cell membrane. One molecule is transported against its electrochemical gradient, whereas the other molecule is transported along its electrochemical gradient. The SGLT2 carrier protein is an example, transporting Na+ with its gradient and glucose against its gradient. This type of secondary transport relies on the Na+/K+ ATPase pump described aboveto ensure that the intracellular Na+ is low enough to pull in additional Na+ and bring glucose along with it.

Pinocytosis

Pinocytosis is mechanism of transport where the cells take up large particles and fluids by "drinking" extracellular fluid. The cellular membrane invaginates to enclose filtrate within a vesicle that pinches

off from the membrane and enters the cell cytoplasm. It then fuses with lysosomes to form endolysosomes where the larger molecules in the fluid are degraded and digested into molecular components.

Receptor-mediated endocytosis

Receptor mediated endocytosis is a variant of pinocytosis; however, in this process a larger protein will bind to a receptor in the membrane before the vesicle forms.

Proximal Tubules

The major function of the proximal tubules (PT) is to reabsorb fluid, electrolytes and nutrients. There is a convoluted portion, which is tortuous and often seen as rings of epithelial cells lining a basement membrane, and a straight portion which is more linear running from the cortex to the medulla. This PT is responsible for roughly 98% of glucose and 65% of sodium reabsorption. The PT is located in the cortex and is composed of simple columnar epithelium joined together by tight junctions. The cell- cell junctions are permeable to small molecules yet permit transport from the tubular lumen into the interstitium in a paracellular manner. The apical brush border is a characteristic and distinguishing feature of the PT. The brush border is composed of microvilli which facilitate reabsorption of small peptides and simple sugars (monosaccharides and disaccharides).

Loop of henle

The **thin descending limb** of Henle travels from the PT and forms a hairpin loop to become the **thin ascending limb** and the **thick ascending limb** before terminating into the distal convoluted tubule. The thin limbs (descending and ascending) are lined by simple squamous epithelium. The thin descending loop is highly permeable to water (which is easily transferred to the interstitium) and less permeable to solutes. On the other hand, the thin ascending limb is highly permeable to sodium but impermeable to water. During this part of the journey, the ultrafiltrate loses water to the interstitium as it travels through the thin descending limb, and the remaining urine becomes hyperosmotic. As the urine travels up the ascending limb, sodium passively diffuses out of the tubular lumen into the interstitium. This creates a hyperosmotic ("salty") medullary interstitium. The thick ascending limb is composed of simple cuboidal epithelium. This portion of the tubule can be distinguished from the PT by lack of a brush border. At this point, the ultrafiltrate has lost enough solutes to become hypoosmotic as it reaches the thick ascending limb.

Distal convoluted tubule

The distal convoluted tubule (DCT) is composed of simple cuboidal epithelium. The DCT is located in the cortex and corticomedullary junction. The functions of this segment include fine-tuning ion concentrations and acid-base balance to maintain homeostasis. The epithelium of the DCT is also involved in reabsorption or excretion of bicarbonate and hydrogen ions to maintain blood pH. It can secrete potassium into the tubular lumen under the influence of the hormone aldosterone and reabsorb calcium in response to parathyroid hormone.

Collecting ducts

The collecting ducts span the renal cortex and medulla and concentrate the urine. The collecting duct is composed of simple cuboidal epithelium. There are two distinct cell types present in the collecting tubules: **intercalated cells** and **principal cells**. The principal cells are the predominant cell type with short sparse microvilli and contain Na+/K+ ATPase pump and aquaporin, which is an anti-diuretic hormone [ADH]-regulated water channel. The intercalated cells are fewer in number, but they are important for the regulation of blood pH. These cells will adjust bicarbonate or hydrogen ion reabsorption.

Papillary ducts

These are continuous with the collecting ducts and their lumens widen slightly as they reach the renal papilla. The epithelium lining these tubules transitions smoothly to the urothelium that lines the renal papilla.

The renal interstitium

The interstitium is composed of fibrocytes and occasional macrophages. The interstitial cells produce the extracellular matrix (collagen and glycosaminoglycans), which surround the nephrons, duct, blood vessels, and lymphatics.

Peritubular capillaries

The vessels that are responsible for reabsorbing all of the molecules that cross the tubular epithelial cells are called peritubular capillaries. In order to do their job efficiently, they need to be immediately adjacent to the tubules. If there is an aggregate of fibroblasts or inflammatory cells, then it is more difficult for them to take up the products delivered to them by the tubular epithelial cells. Likewise, this intricate tortuous network of capillaries delivers oxygen and glucose to the tubules as well. Loss of the peritubular capillaries or an increase in the distance between the tubules and the capillaries will therefore impact tubular epithelial energy balance and functionality.

Vasa recta

In addition to the peritubular capillaries that surround the cortical tubules, there is another aggregate of thin-walled vessels known as the vasa recta. These vessels run alongside the Loop of Henle.

The Juxtaglomerular Apparatus

The juxtaglomerular apparatus (JGA) controls the rate of the glomerular blood flow depending on various factors such as hydration, blood volume and ion concentration. The JGA includes the **macula densa** and **juxtaglomerular cells**. The macula densa is a short segment of the distal convoluted tubule that is located directly adjacent to the afferent and efferent arterioles at the vascular pole of the glomerulus. The juxtaglomerular cells are within the interstitium between the macula densa and the vascular pole. The macula densa cells sense sodium chloride concentration in the lumen and stimulates the juxtaglomerular cells to synthesize and release renin into the bloodstream. This happens in states of hypovolemia or low sodium chloride concentration in the blood (and in the urinary filtrate). This activates the renin-angiotensin-aldosterone system (RAAS) and triggers several key important steps to help improve blood pressure. The release of renin in the bloodstream causes hydrolysis of angiontensinogen to form angiotensin I. Angiotensin I is converted by angiotensin-converting enzyme (ACE) to angiotensin II. Angiotensin II triggers the adrenal glands to produce and release aldosterone, whichacts on tubules to upregulate sodium and water reabsorption back into the blood.

Renal blood supply

The cortex of the kidney receives the majority of renal blood flow. The renal artery enters at the hilum of the kidneys and branches into the interlobar artery. The interlobar artery then travels between the pyramids and continues as arcuate artery, which arches along the corticomedullary junction. The interlobular artery is a stem off of the arcuate artery and travels through the cortex towards the capsule. As the interlobular artery ascends towards the cortex, branches of afferent arterioles are sent to each glomerulus. The afferent arterioles give rise to the capillaries that form the glomerulus. The glomerular capillaries reunite and form the efferent arteriole which exits at the vascular pole. The efferent arterioles of cortical nephrons give rise to the peritubular capillaries. As discussed previously, in the juxtamedullary nephrons, the peritubular capillaries form the ascending and descending vasa recta. The peritubular capillaries and vasa recta then drain into the interlobular vein, which unites with others to form the arcuate vein. Then the interlobular vein exits the kidney as the renal vein.

Urine elimination

Once urine passes through the collecting ducts and enters the renal papilla, it will be delivered by the ureters to the urinary bladder. Urothelium lines the ureters, bladder, and the urethra. The urothelium is also referred to as transitional epithelium because it can transition from a plump rounded epithelium in a contracted bladder to a flattened epithelium in a distended urine-filled bladder. The urothelium is enveloped by a lamina propria and smooth muscle that is arranged in two layers (inner longitudinal and outer circular). Urinary bladder innervation it is under control of both the parasympathetic and sympathetic nervous system. Contraction of the urinary bladder occurs with the parasympathetic system via the pelvic nerve. The sympathetic nervous system relaxes the bladder smooth muscle and contracts the urethral smooth muscle (to keep urine from leaking out) via the hypogastric nerve.

Chapter-12

Male reproductive system

This chapter will cover the male reproductive system.

Chapter Learning Objectives

By the end of this chapter, you should be able to do the following:

- Identify the primary factor behind sex determination.
- Identify which embryonic structures form the tubular genitalia for males and females.
- List the periods or stages of spermatogenesis and the cell types associated with them.
- Describe the process of spermiogenesis.
- Outline the pathway that sperm travels from the testicle to the urethra.
- List the accessory sex glands and list which species have which glands.

Review Questions

By the end of this chapter, you should be able to answer the following:

- How does sex determination take place in the male?
- Define the locations of the following cells in the testicular parenchyma: Leydig cell, Sertoli cell, Type A spermatogonia, spermatocyte, spermatid. What are their functions?
- What are the structures that form a mature sperm cell?
- What are the structures which form the male tubular genitalia? What are they lined by?
- List the accessory sex glands and described their locations. Which species have what glands?
- What is the difference between the corpora cavernosa and the corpora spongiosum?
- How does the penis form?

Overview

The male reproductive tract functions to propagate germ cells by development of haploid spermatozoa in the testes and delivery of the spermatozoa to the female reproductive tract. The male reproductive tract consists of the gonads (testes), the tubular genitalia (epididymis, ductus deferens and urethra) and the external genitalia (penis and scrotum).

Development and structure of the gonads and tubular genitalia

Gonadal development

The first step in gonadal development takes place when the primordial germ cells migrate from the allantois (portion of the placenta) to the genital ridge, the structure that will become the undifferentiated gonad. The gonadal ridges are protuberances within the coelomic cavity of the developing fetus. The structures of the primitive gonad that are **not** composed of primordial germs cells (i.e. somatic gonadal cells) are derived from local mesenchymal cells, coelomic epithelium and cells derived from mesonephric tubules. Ultimately Sex determination of the undifferentiated gonad is dependent on the presence of the sex determining region Y (SRY) gene found on the short arm of the Y chromosome. Therefore without the Y chromosome and SRY gene, differentiation to a female gonad will take place.

Development of tubular genitalia

The formation of male tubular genitalia is directed by the hormones secreted by the developing testes. The secretion of testosterone from the Leydig cells and paramesonephric inhibitory hormone from the Sertoli cells induce the differentiation of the Wolffian body (mesonephros) and the Wolffian ducts (mesonephric ducts). Paramesonephric inhibitory hormone inhibits the development of the paramesonephric ducts (Muellarian ducts). In the absence of developing testes and the presence of estrogens, the paramesonephric ducts develop into the uterus, uterine tubes and cranial vagina. Estrogen also stimulates the development of female external genitalia, the caudal vagina, the vestibule and the clitoris (more in chapter 13).

The testes

The testes have two functions. The first is to produce spermatozoa and the second is to function as an endocrine organ. A portion of mesonephric cells develop into cords of tissue called seminiferous cords which incorporate the migrating primitive germ cells into their structure. The seminiferous cords, also known as medullary cords, develop eventually into Sertoli cells, while mesodermal cells in between seminiferous cords develop into Leydig cells (interstitial cells). Sertoli cells secrete factors that regulated spermatogenesis and spermiogenesis, the function of Leydig cells. They also secrete tubular fluid and produce inhibin that regulates hormone production. Leydig cells are responsible for testosterone production.

Spematogenesis

Germ cells residing in the developing testes are referred to as spermatogonia. Successive rounds of mitotic division increase the number of cells that can end as the terminally differentiated spermatogonia, known as spermatozoa. Spermatogenesis is quiescent until puberty and the process takes place within the seminiferous tubules.

The spermatogenic process can be divided into the mitotic, meiotic and metamorphic periods. Spermatogenesis takes places at different stages throughout the length of the seminiferous tubules.

The mitotic phase of spermatogenesis results in increased numbers of spermatogonia, allowing for an enlarged replicative pool before the first meiotic division. Two types of spermatogonia are present. Type A spermatogonia have a large round nucleus with condensed chromatin, peripheral nucleoli and a nuclear vacuole occasionally. These spermatogonia maintain the replicative pool of germ cells. Type B spermatogonia have dispersed chromatin, central nucleoli and no nuclear vacuole. These spermatogonia eventually produce spermatozoa.

Formation of spermatocytes represents the meiotic stage of spermatogenesis. Primary spermatocytes are derived from the spermatogonium and are still diploid cells. Primary spermatocytes undergo the first meiotic division to become haploid cells called secondary spermatocytes. These soon enter the second meiotic division to become spermatids.

Spermiogenesis

The conversion of spermatids to spermatozoa represents the metamorphic stage known as spermiogenesis. During this process, granules in the Golgi vesicles coalesce to form a large vesicle (acrosomal vesicle) containing a structure called the acrosome. The acrosomal vesicle expands to cover half of the nucleus. The acrosome and the acrosomal vesicle form the (acrosomal) head cap. Simultaneously, centrioles, mitochondria and the Golgi remnants migrate to the cytoplasm opposite to the acrosomal cap. One centriole elongates to become the developing flagella. The centrioles also interface with the nucleus to form the neck. During the entire process, the spermatid elongates and it nucleus condenses.

The mature spermatozoon is divided into 5 sections

1. Head
2. Neck – where the centriole migrated and connects the head to the tail
3. Midpiece
4. Principal piece
5. Tail or end piece

The stage of spermatozoon development is variable between regions of seminiferous tubules and progression is cyclical in nature. This variability is known as the spermatogenic wave.

The activity in the seminiferous tubules is primarily regulated by the products of the hypothalamus and the anterior pituitary. Luteinizing hormone (LH) and Follicle stimulating hormone (FSH) are secreted by the anterior pituitary. FSH initiates spermatogenesis, while LH stimulates testosterone secretion that is required for spermatozoon maturation.

Testicular structure

A capsule composed of dense collagenous material that encloses the testicle is called the tunica albuginea. Most species contain a vascular layer within the tunica albuginea. Horses often contain smooth muscle fibers within the tunica albuginea. The double layer of simple squamous epithelium and mesentery directly covering the testes and apposed to the tunica albuginea is the tunica vaginalis. The parenchyma of the testicle is divided into lobules by loose connective tissue bands (septuli testes). These lobules are composed of tubules lined by stratified epithelium composed of maturing germ cells (spermatogonia, spermatocytes, spermatids and spermatozoa) and Sertoli cells. These tubules are supported by a basement membrane which contains fibroblasts and myofibroblasts. The Sertoli cells are triangular or oval in shape with a prominent nucleolus and fine chromatin. These tubules (seminiferous tubules) constitute the exocrine portion of the testes. The maturing germ cells are stratified so that the spermatogonia constitute the basilar compartments closest to the basement membrane. The spermatocytes, spermatids and spermatozoa constitute the next layer or the apical compartment with the spermatocytes adjacent to the basilar compartment and the spermatozoa closest to the lumen.

The endocrine portion of the testes is comprised by the Leydig or interstitial cells which are present between seminiferous tubules primarily located in the extra-tubular connective tissue. These cells are responsible for testosterone secretion.

As spermatozoa mature in the seminiferous tubules, they are guided out of the testes by the rete testis (randomly arranged) or tubuli recti (straight tubules) which are connected to the seminiferous tubules. These tubules eventually lead to the extragonadal tubular structures.

Male Tubular Genitalia

These structures are derived from the caudal remnants of embryonic structures called the mesonephric ducts (Wolffian ducts). They function to transport spermatozoa towards the urethra.

The structures that are derived from the mesonephric ducts include:

1. **Efferent ductules** – These structures connect the rete testes to the ductus epididymis. They consist of 6-20 coiled tubules lined by ciliated columnar epithelium which assist in the movement of spermatozoa.
2. **Ductus epididymus** – This coiled tube functions to store spermatozoa. Some maturation of spermatozoa also takes place here. The epididymus and its associated connective tissue and muscle are divided into the head (where the efferent ductules terminate), the body and the tail. The tail is continuous with the ductus deferens. The mucosa of the epididymus is lined by ciliated pseudostratified columnar epithelium. Under the lamina propria lies a thin band of smooth muscle (muscularis mucosa) which is slightly thicker at the tail. The submucosa of this structure is continuous with the connective tissue of the tunica albuginea.
3. **Ductus (vas) deferens** – This structure is lined by pseudostratified columnar epithelium. In contrast to the previous tubular structures, the lamina propria lies on top of a thick smooth muscle tunica muscularis which is divided into inner and outer layers. The ductus deferens contains an outer serosa.

Male Accessory Sex Glands

The accessory sex glands function to: 1. nourish spermatozoa, 2. activate spermatozoa, 3. clear the urethral tract before ejaculation, 4. produce secretions that assist in transport to the female reproductive tract, 5. in some species (rodents), form a plug that helps retain spermatozoa in the female reproductive tract.

There is variation between species in terms of presence, size, location and function of these glands. The glands are derived from outpouchings of the ductus deferens or the urethra during development.

The structure of most of the accessory sex glands is arranged as branched tubular or tubuloalveolar pattern which are arranged in lobules. Each lobular structure drains into the ductus deferens or urethra.

The four major accessory sex glands are the ampulla, the vesicular glands (seminal vesicles) and the bulbourethral glands.

1. Ampulla – This gland is an enlargement of the terminal portion of the ductus deferens and consists of branched tubular or tubuloalveolar glands without a specialized excretory duct. The ampulla is fully developed in ruminants, and horses where the glandular structures are located in the lamina propria. These glands are absent in dogs, cats and swine.
2. **Vesicular glands (seminal vesicles)** – These glands are typically paired and are composed of simple columnar glandular epithelium which is arranged in lobules. The excretory ducts are lined by stratified columnar epithelium. Vesicular glands are absent in carnivores but present in horses, swine and ruminants.
3. **Prostate glands** – The prostate glands consists of the main body and a disseminate portion. The body is surrounded by a thick collagenous capsule while the disseminate portion is embedded in loose collagenous matrix extending along the dorsal aspect of the urethra and eventually extends laterally and ventrally to join the body. The gland is histologically composed of tubuloalveolar structures composed of low columnar to cuboidal cells, have apocrine gland type secretory activity and have acidophilic granules and lipid droplets in the epithelial cytoplasm. The body is well developed in carnivores and horses while the disseminate portion is better developed in cattle and swine. Rams do not have a well developed prostate.
4. **Bulbourethral glands** – These are paired structures located dorsolaterally to the pelvic urethra and are composed of paired tubuloalveolar glands with columnar epithelium. The ducts of this gland are lined by columnar, pseudostratified or transitional epithelium. All domestic animals except for dogs have bulbourethral glands. The mucus production from this gland has lubricating function and clears the urethra of urine.

Male Genital Ligaments

The gubernaculum develops during testicular development as a column of fibrous connective tissue between the inguinal regions to the caudal pole of the testis. The abdominal component of the gubernaculum is covered by mesentery. The extra-abdominal component is initially attached to the abdominal wall, but moves caudally as the inguinal rings and scrotal sac develop. During late gestation or the early post-natal period (depending on the species), the intraabdominal developing testes descend through the inguinal rings to the scrotum while a portion of the gubernaculum shortens and eventually becomes the scrotal ligament. The mechanism in which the gubernaculum condenses and shortens is not completely understood.

Male External Genitalia

The major external structures are the penis, penile urethra and scrotum. Embryologically, the cloacal folds fuse ventrally to form the genital tubercle. The genital tubercles elongate and the accompanying urogenital folds fuse to convert the groove in between the folds to a tube which becomes the urethra. The surrounding genital tubercles form the penis. Proliferating tissue lateral to the urogenital folds called the genital swellings form the scrotum. The midline region of fusion between the genital swellings becomes the medial raphe which separates the testes. The prepuce is formed from loose connective surrounding the penis.

The urethra is divided into the pelvic and penile portions. Both portions are lined by transitional epithelium. Under the lamina propria, portions of the tunica muscularis is formed from striated muscle rather than smooth muscle. A tunica adventitia is present in this region. The mucosa of the penile urethra changes to squamous epithelium at the external opening. Glands are present in the lamina propria, particularly in the stallion and boar. The tunica muscularis is composed of smooth muscle in the penile urethra.

The anatomy of the penis can vary tremendously by tissue composition and organization between species. In general terms, the penis can be divided into the glans, the body and the roots. The glans is covered by a reflection of preputial skin. The tissue of the glans can be composed of vascular erectile tissue, cartilage in bulls or in the case of the carnivores, bony tissue. The body is enclosed by a capsule which is thick and prominent in swine and ruminants (fibrous penis) and thin in carnivores (vascular penis). The subcapsular tissue encloses vascular erectile tissue (corpora cavernosa and corpus spongiosum). This tissue is composed of collagenous tissue and vascular spaces which can expand due to the presence of elastin fibers in the connective tissue. The roots are the most proximal aspects of the penis and are similar in structure to the body.

The scrotum is enclosed in haired skin with an underlying layer of smooth muscle (tunica dartos). The scrotum contains invaginations of peritoneal mesentery (the parietal and visceral vaginal tunics) which enclose the testes, epididymus, ductus deferens and associated vascular structures.

Chapter-13

Female Reproductive System

This chapter will cover the female reproductive system.

Chapter Learning Objectives

By the end of this chapter, you should be able to do the following:

- ❖ Identify the primary factor behind sex determination.
- ❖ Identify which embryonic structures form the tubular genitalia for males and females.
- ❖ Compare and contrast the histological structure of each stage of follicular development.
- ❖ Identify the differences between ovarian structure in different species.
- ❖ List the layers forming the following structures: Uterine tube, uterus, cervix, vagina, and vestibule. What the composition of the innermost aspect of these structures.
- ❖ Contrast the cytologic changes in vaginal mucosa during each stage of the estrus cycle.
- ❖ Outline the structures forming the avian reproductive tract.
- ❖ Describe the structure of the mammary gland.

Review Questions

By the end of this chapter, you should be able to answer the following:

- ❖ What's the difference between a primary oocyte and a primary follicle?
- ❖ What's the difference between a secondary oocyte and a secondary follicle?
- ❖ List the structural differences between the primary, secondary and tertiary follicle.
- ❖ Identify the following structures: Granulosa cells, theca interna, theca externa, cumulus oophorus, and corona radiata.
- ❖ What are the segments of the female genital tract?
- ❖ What are the segments of the uterine tube? What is the order from ovary to uterus?
- ❖ What are the layers of the uterus? What are the individual layers of the endometrium called?
- ❖ List (in all species) the types of surface epithelium which lines the: uterine tube, uterus, cervix, and vagina.

- ❖ What are the species differences in cervical structure?
- ❖ What are the various changes which take place in the canine vaginal mucosa during estrus?
- ❖ List the structures in the female avian reproductive tract.

Overview

The female reproductive system is comprised of the gonads (ovaries), the tubular reproductive tract (uterine tubes, uterus, cervix, vagina and vestibule) and the external genitalia. While the mammary glands are not part of the female reproductive tract, it is a tissue integral to the reproductive biology of the female. Learning histology of the female reproductive system can be difficult as tissues can change in morphology in response to hormonal stimulation. But understanding the structure of the tissue greatly assists in understanding the estrus cycle in domestic animals.

Gonadal Development

The first step in gonadal development takes place when the primordial germ cells migrate from the allantois (portion of the placenta) to the genital ridge, the structure that will become the undifferentiated gonad. The gonadal ridges are protuberances within the coelomic cavity of the developing fetus. The structures of the primitive gonad that are **not** composed of primordial germs cells (i.e. somatic gonadal cells) are derived from local mesenchymal cells, coelomic epithelium and cells derived from mesonephric tubules. Ultimately Sex determination of the undifferentiated gonad is dependent on the presence of the sex determining region Y (SRY) gene found on the short arm of the Y chromosome. Therefore without the Y chromosome and SRY gene, differentiation to a female gonad will take place.

Development of Tubular Genitalia

The formation of male tubular genitalia is directed by the hormones secreted by the developing testes. The secretion of testosterone from the Leydig cells and paramesonephric inhibitory hormone from the Sertoli cells induce the differentiation of the Wolffian body (mesonephros) and the Wolffian ducts (mesonephric ducts) (refer back to chapter 12). Paramesonephric inhibitory hormone inhibits the development of the paramesonephric ducts (Muellarian ducts). In the absence of developing testes and the presence of estrogens, the paramesonephric ducts develop into the uterus, uterine tubes and cranial vagina. Estrogen also stimulates the development of female external genitalia, the caudal vagina, the vestibule and the clitoris.

The Ovary

Ovarian function and development

The ovaries function 1.) to produce ova, 2.) as an endocrine gland (i.e. secrete mediators into the bloodstream to act at a distant site). During ovarian development from the primordial gonad, structures termed sex cords form which are likely derived from mesothelial cells forming the gonads. These linear plates of tissue incorporate the migrating germ cells into their structure.

Oogenesis refers to the development of oocytes within the ovary. The process can be divided into three stages – fetal, prepuberty (resting stage) and puberty.

During fetal oogenesis, the sex cords break down and the germ cells undergo enhanced mitosis. In most mammals, this prolonged mitotic activity ceases before or soon after birth. These cells arrest in prophase of meiosis I (primary oocytes), are enclosed in basal lamina surrounded by follicular cells and are referred to as primordial follicles.

The following concepts must be understood regarding oogonia and oocytes in relationship to the prepubertal, pubertal and post pubertal periods.

1. Primary oocytes enter a resting phase in which full maturation does not occur until stimulation by gonadotrophic hormones takes place (puberty).
2. Large numbers of post mitotic germ cells (oogonia) and primary oocytes undergo atresia or death in the prenatal period, the prepuberty period and the post puberty period.

Following puberty, the primary oocyte surrounded by follicular cells enters follicular maturation when stimulated by gonadotrophic hormones. This maturation occurs in cyclical waves. In regards to the developing follicle, the surrounding follicular cells proliferate, become stratified and change morphology from a spindle shaped cell to a cuboidal phenotype (called granulosa cells).

Completion of the first meiotic division results in the formation of the secondary oocyte and extrusion of the first polar body. This change typically takes place before ovulation, except in dogs, where formation of the secondary oocyte takes place after ovulation. The second meiotic division takes place during fertilization and results in the extrusion of the second polar body.

Ovarian structure and histologic organization

The ovary is divided anatomically into the cortex and medulla. The cortical aspect of the ovary is covered by cuboidal epithelium during development that converts to squamous epithelium with age. The cortical parenchyma is composed of follicles (oocytes and follicular cells), interstitial cells and collagenous connective tissue stroma. The ovarian medulla contains large arteries and veins, lymphatics, nerves embedded in a loose collagenous matrix. The rete ovarii are also present in the medulla, which are cords of cells found in the medulla homologous to the rete testis.

In most domestic species, the cortex contains the developing follicles. However in the horse, the anatomy of the ovary is reversed. In this species, the follicular development takes place in the region of the ovary corresponding to the medulla. A concavity forms in the ovary called the ovulation fossa, corresponding to the site in the equine ovary which ovulation occurs. In addition, ovaries in the equine fetus and neonates are markedly enlarged compared to the size of the animal. This enlargement in attributed primarily to hyperplasia and hypertrophy of the interstitial cells. Subsequent atrophy takes place post-natally.

Follicular development

Development of the ovarian follicle is a sequential process which is primarily directed by the influence of gonadotrophins (follicle stimulating hormone [FSH] and luteinizing hormone [LH]) and can be variable between species.

The sequence of follicular development is 1.) primordial follicle, 2.) primary follicle, 3.) secondary follicle, 4.) tertiary follicles.

- **Primordial follicles** contain a primary oocyte and are surrounded by a single layer of flattened follicular cells.
- **Primary follicles** still contain a primary oocyte but the follicular cells become more cuboidal and are now known as granulosa cells. Follicular (granulosa) cells proliferate (membrane granulosa) but are separated from the oocyte by a thick periodic acid Schiff (PAS) positive basement membrane called the zona pellucida. The organized stromal cells around the follicles are called theca cells.
- **Secondary follicles** start develop spaces between granulosa cells that coalesce to eventually form a large space called the follicular antrum. The granulosa cells secrete PAS positive material into these spaces. The stromal cells surrounding the follicle form two layers, the theca interna and the theca externa.
- **Graafian (tertiary, mature) follicles** are large preovulatory follicles which bulge from the surface of the ovary. Once the follicular antrum is formed, the oocyte is surrounded by a remnant of granulosa cells called the cumulus oophorus. The cells of the cumulus oophorus immediately adjacent to the oocyte are known as the corona radiata.

Endpoints of follicular development

A follicle only has two potential fates, atresia or ovulation. Follicular atresia or degeneration of the developing follicle can take place any time during development. During this process the oocyte liquefies with subsequent collapse of the zona pellucida and the wall of the follicle. Fibrous tissue, phagocytes and vascular tissue replace the region of the atretic follicle while the remnants of the theca interna undergo hypertrophy. Ovulation involves the rupture of the follicle and release of the oocyte. This is induced by a surge of luteinizing hormone released by the anterior pituitary. Only the corona radiata is present on the oocyte after release from the follicle.

Consequences of ovulation

The corpus hemorrhagicum is formed soon after ovulation from hemorrhage into the remnants of the follicular structure resulting in a large blood clot. The remaining granulosa cells and cells of the theca interna undergo hypertrophy, proliferate and invade the corpus hemorrhagicum, transforming into granulosa lutein cells and theca lutein cells respectively (luteinization). These cells organize into a highly vascular glandular organ called the corpus luteum (yellow body). The yellow color of this structure is imparted by intracellular accumulations of a pigment, lutein, in some species (equine, bovine, canine and feline). In ewes and sows, other lipid pigments impart the yellow color. If fertilization successfully takes place, the corpus luteum will be maintained. However, if fertilization does not take place, the corpus luteum will degenerate and be replaced by fibrous connective tissue forming the corpus albicans. The corpus luteum is the main source of progesterone, which is important for the maintenance of pregnancy and synergistic with estrogen during estrus.

Female Tubular Genitalia

Female tubular genitalia is composed of the uterine tubes (oviducts), the uterine horns, the uterine body, the cervix and the vagina.

The uterine tubes are projections from each of the uterine horns which interface with each ovary. These structures provide transport of the oocyte towards the uterine horns and provide a site for fertilization. The uterine tubes are divided anatomically into the infundibulum, the ampulla and the isthmus. The mucosa is typically composed of ciliated columnar epithelium although in some species (bovine and porcine) it is composed of ciliated pseudostratified epithelium. The cilia move the fertilized ova towards the uterine horns.

The uterus is responsible for the maintenance of pregnancy and the transport of semen, although in some species, such as the horse, it is also the receptacle for semen. In other species, semen is deposited in the cervix or vagina. The shape of the uterus is highly species dependent. The bicornuate uterus has two large uterine horns, a uterine body and a single cervix – porcine, canine and feline, ruminants and equine. However, between species, the bicornuate uterus varies in structure. In canines, felines and swine, the uterine horns are highly developed. In cattle, sheep and horses, the uterine horns are poorly developed.

Rodents and rabbits have a duplex uterus that has two uterine openings with corresponding cervices which open into a common vagina.

The cross sectional anatomy of the uterus is divided into the endometrium, the myometrium and the perimetrium. The endometrium is typically covered with simple columnar epithelium although in some species, pseudostratified or cuboidal epithelium can be present as well. This epithelium rests on a loose collagenous tissue called the lamina propria. Beneath the mucosa, simple or branched tubular uterine glands extend from the lamina propria to the submucosa.

The endometrium is divided histologically into 3 layers.

1. **Stratum basalis** – Deepest layer adjacent to the myometrium.
2. **Stratum spongiosum** – Broad intermediate layer with a spongy appearance.
3. **Stratum compactum** – Superficial layer with a compact stromal appearance.

Some species variation exists between the distribution of glands in the endometrium. The ruminant endometrium contains large vascularized protrusions (caruncles) which have relatively little glandular tissue. In addition, species variation exists in the degree of coiling at the distal ends of the glands. The depth of the gland penetration into the lamina propria and submucosa is also dependent on the stage of the estrus cycle.

The myometrium is analogous to the muscular layers found in the small and large intestine. It is composed of smooth muscle arranged in an inner circular layer and an outer longitudinal layer. Between the smooth muscle layers is a prominent vascular layer.

The perimetrium is the outermost layer and is composed of loose connective tissue, lymphatics and small arteries and veins with an overlying mesothelial surface.

The cervix lies between the uterine body and the vaginal vault where it acts as a barrier during certain phases of the estrus cycle and during pregnancy.

The mucosa of the cervix is thrown up in longitudinal folds and its mucosal composition varies between species. In the canine, the mucosa is composed of stratified squamous epithelium. The bovine cervix is contains large numbers of goblet like cells with interspersed ciliated epithelium. The structure of the cervix exhibits species variability as well. The bovine cervical mucosa contains ring- like folds which span its circumference. Longitudinal ridges are present which intersect the rings. Small ruminants (sheep and goats) have more prominent and numerous ridges which make uterine catheterization more difficult. The equine cervix is shorter (~ 6 cm) and has mucosal folds which are prominent at the caudal aspect but disappear cranially. The canine cervix is also short and has a dorsal fold of mucosa which extends into the vagina as well as transverse grooves. The porcine cervix is long (~ 25 cm) and has numerous interdigitating mucosal prominences that occlude the lumen.

The vagina is composed of four layers, the mucosa, the lamina propria, the muscularis and the adventitia. It is formed from the embryonic vaginal plate and the fused ends of the paramesonephric ducts. In the cranial vagina, the adventitia is replaced by a serosa. The mucosa in all species, except the cow, is composed of stratified squamous epithelium. The bovine vagina is lined by stratified columnar epithelium with interspersed goblet cells. Initially, the lumen of the vagina is separated from the vestibule by a thin membrane called the hymen.

The vaginal mucosa undergoes cyclical variation depending on the stage of the estrus cycle and this is of primary value in the canine. By examining the cytology from the vaginal vault, the approximate stage of estrus can be predicted.

1. During **anestrus** (period of inactivity), the vagina contains large numbers of non-cornified, round to oval epithelial cells with large uniform nuclei. A few neutrophils may be present.
2. During **proestrus**, large cornifying epithelial cells with shrunken or absent nuclei are present. Neutrophils are rare. Erythrocytes and bacteria may be present.
3. During **estrus**, large cornified epithelial cells predominate. Many cells will not have nuclei. Erythrocytes are reduced.
4. During **diestrus**, the smaller non-cornified epithelial cells reappear and neutrophils are dominant. Erythrocytes are rare.

Female Genital Ligaments

As the mesonephros degenerates (since it is a precursor to male tubular genitalia), the developing ovaries and ductular genitalia is enveloped in folds of peritoneum. This structure is called the genital fold and its divisions are named by the attached portions of the reproductive tract. The mesovarium supports and suspends the ovaries in the peritoneum. The mesosalpinx attaches to the uterine tube and portions of the ovary. The mesometrium (broad ligament of the uterus) supports the uterus in the pelvic canal and abdomen.

Female External Genitalia

The vestibule forms from the urogenital sinus and, as mentioned previously, it is separated from the vagina initially by the hymen. The vestibule develops as a structure separate from the vagina and fuses with it later in development. The vestibule, however, is similar in structure to the vagina and is lined by stratified squamous epithelium. Below the lamina propria of this structure is a tunica muscularis with an inner circular and outer longitudinal smooth muscle layer. The outer muscular layer is composed of two different muscles, the constrictor vestibule and the constrictor vulvae.The vestibular glands are tubuloalveolar glands present in the constrictor vestibule muscle and are homologous to the male bulbourethral glands and the urethral glands.

The clitoris is homologous to the penis. It is composed of the body, glans and preputial covering. The body is composed of cavernous tissue, smooth muscle and adipose tissue surrounded by a thick collagenous capsule. The glans is composed of cavernous tissue (in canines and horses) or highly vascular connective tissue. The preputial covering is the reflection of the stratified squamous epithelial lined vestibular mucosa.

The labia are simply folds of skin just caudal to the opening of the vestibule. The histologic structure is similar to what you will read in the integumentary chapter.

Avian Female Reproductive Tract

In contrast to most mammalian reproductive systems, the avian reproductive tract allows internal fertilization, but allows survival and development of the ovum outside of the body. Only the left ovary and oviduct are retained in the adult female animal. The avian ovary is not as compact as the mammalian ovary, but it does have a cortex and a poorly defined medulla. Ovarian follicles can reach a diameter of 30 mm, contain a single oocyte with a single layer of granulosa cells and a think theca interna and externa. A corpus luteum is not formed after ovulation.

The oviduct is divided into the infundibulum, the magnum, the isthmus and the shell gland. The infundibulum is the extension of the oviduct with a vascularized mucosa lined by pseudostratified columnar. The magnum deposits the egg white and is lined by ciliated and nonciliated columnar cells. The lamina propria contains branched tubular glands. The isthmus is responsible for formation of the shell membranes and is similar in structure to the magnum. The shell gland is essentially the avian uterus but does not allow for ova implantation. This organ is lined by pseudostratified columnar epithelium with sporadic ciliated cells. The avian vagina is similar in histologic structure to the shell gland and contains glands which store sperm in preparation for fertilization.

The Mammary Gland

The mammary gland functions primarily to produce milk and colostrum for neonates. However, there are many anatomic differences between the mammary glands of different species. This organ is composed of compound tubuloalveolar glands. The secretory component of these glands has two parts. The first part is composed of columnar cells that form the alveolar epithelium. The second part is composed of columnar cells forming the intralobular ducts. The cells typically exhibit apocrine secretion. Inactive secretory epithelium often has a lower profile or more cuboidal appearance. The space between alveoli and ducts is taken up by loose, well vascularized connective tissue. Myoepithelial cells or epithelial cells with contractile properties are present around alveolar structures. Sloughing epithelial cells, macrophages and much smaller numbers of other leukocytes are also seen. Concretions of casein and cellular debris also known as corpora amylacea can be seen in glandular structures.

One or two alveoli drain into each intralobular duct. The intralobular ducts drain into a lobular duct and several lobular ducts drain into and form a lobar duct. Each lobar duct drains a lobe of the mammary gland. Many lobar ducts drain into the lactiferous sinus, the common chamber at the base of each gland. The sinus is constricted at its distal end (annulus) where the sinus enters the teat structure. The sinus narrows to the lactiferous sinus also known as the teat canal. While most of the lactiferous sinus is lined by stratified columnar epithelium, the canal is lined by stratified squamous epithelium.

In lactating mammary glands, the number of alveoli is far greater than the numbers in inactive mammary glands. The inactive glands have large amounts of glandular tissue replaced by fibrous connective tissue and adipose. The larger ductal elements remain intact in these glands.

Chapter-14

The Eye

The eye itself can be divided into 3 concentric tunics plus the internal components. This chapter will cover the eye.

Chapter Learning Objectives

By the end of this chapter, you should be able to do the following:

- Describe the three concentric tunics of the eye
- Describe the components of each at the level of the retina, lens and iris, and cornea.

Review Questions

By the end of this chapter, you should be able to answer the following:

- What are the layers of the cornea? Explain the function of each layer and how it responds to injury/heals.
- How does the tear film work and relate to the corneal epithelium? 3
- What structures comprise the vascular tunic of the eye? What is the specific function of each of the regions of the vascular tunic?
- Where is the tapetum? What is its function?
- Describe the anatomy/ histology of the lens and the regional variations within the lens.
- Describe the 10 layers of the retina, and how vision occurs as a photon of light enters the eye.
- Trace the path of light, describe how each layer of the eye affects this photon and how it is ultimately sent to the brain.

Overview

The eye itself can be divided into 3 concentric tunics plus the internal components. The three tunics from the outside surface of the eye inward are, (1) the fibrous tunic (cornea and sclera), (2) the vascular tunic (iris, ciliary body, and choroid) and (3) the neuroectodermal (nervous) tunic (retina). These tunics inwardly decrease in size as they become more specialized. Internally, the eye contains the lens and the aqueous and vitreous humors.

Below is a breakdown of the eye structure:

3 Concentric Tunics

1. **Fibrous Tunic**

   Cornea Sclera

2. **Vascular Tunic**

   Iris

   Ciliary Body Choroid

3. **Nervous Tunic**

   Retina

   Additional Structures

   Lens

   Aqueous Humor

   Vitreous Humor

   Optic Nerve

Fibrous Tunic

The fibrous tunic is comprised of 2 components, the clear **cornea** and the opaque **sclera**. They differ primarily in the arrangement or organization of their collagen and water content with this accounting for the difference in their transparency.

Cornea

The mammalian cornea is a transparent, avascular structure that functions to transmit and refract light, and as a protective barrier for the internal ocular contents. Although the cornea is avascular, it does contain nerve fibers. These are branches of the ophthalmic branch of cranial nerve 5 and are primarily concentrated in the anterior third of the cornea.

In general, the cornea of various species is remarkably similar, differing in thickness, curvature, and shape only. The cornea is thickest peripherally. The histologic layers of the cornea, from the outside in, are the corneal epithelium, the corneal stroma, Descemet's membrane and corneal endothelium.

Corneal Epithelium

The corneal epithelium is composed of a stratified squamous, non-keratinized cell layer. Embryologically, the cells of the epithelium originate from surface ectoderm. The number of epithelial cell layers varies in thickness with the species and increases in number towards the limbus. Epithelial cells are anchored to each other by desmosomes and to the basement membrane by hemidesmosomes and anchoring fibrils. External to the epithelium is the tear film. The tear film supplies oxygen to the cornea and approximately 10% of the glucose used by the cornea. The epithelial cells are divided into basal, wing, and superficial cells. The superficial cells have microvilli, microplicae, and a glycocalyx on their surface to stabilize the tear film. Beneath the superficial cells are the wing cells that represent a transition zone between the basal and superficial cells. A single layer of tall, columnar basal cells is the deepest layer of the epithelium. These cells contain abundant glycogen stores and are the source of cells for the renewal of the outer layers. The turnover rate of these cells is 7 days. In addition to the outer layers of the epithelium, the basal cell layer is also responsible for the secretion of the basement membrane. This membrane is composed of type IV collagen and glycosaminoglycans (GAG's).

Stroma

The corneal stroma comprises 90% of the corneal thickness. It is composed of collagen fibers, keratocytes, and GAGs that account for 22% of the stroma. The remaining 78% of the stroma is water. The collagen fibers are regularly arranged in lamellar sheets and span the entire diameter of the cornea. These lamellae lie parallel to the corneal surface, but obliquely to adjacent lamellae. The transparency of the cornea depends on the arrangement of these lamellae and the distance between them. It is hypothesized that an interfiber distance of less than a wavelength of light allows transmission with minimal interference. The normal collagen in the corneal stroma is type I and is more heavily glycosylated than in skin. When injured this is replaced by type III collagen. Between the corneal collagen fibers are flat cells termed keratocytes and GAGs. The keratocytes are capable of fibroblast transformation and phagocytosis, but normally are a quiescent population of cells. The GAGs of the stroma affect hydration, thickness, and transparency. Lymphocytes and Langerhans cells are also reported in the cornea, especially near the limbus.

Descemet's membrane

The posterior limiting membrane or Descemet's membrane is a basal lamina secreted by the endothelium . It is comprised of type IV collagen and is produced throughout life, continuing to increase in thickness with age. It is insoluble except in strong acid or alkali and is more resistant to collagenase than is the corneal stroma.

Endothelium

The innermost layer of the cornea is the endothelium. The origin of this layer is thought to be the neural crest cells. The endothelium is a single cell layer of hexagonal cells, 15-20 um in diameter, attached to each other by terminal bars. In most species these cells are capable of mitosis only very early in life. The greatest cell density of the endothelial cells is in the neonate. Once damaged, endothelial cells repair defects through enlargement and spreading rather than by replacement.

Sclera

The sclera constitutes the major portion of the fibrous tunic and joins the cornea at the limbus. Anteriorly it is covered by bulbar conjunctiva beneath which is the episclera, the scleral stroma, and the lamina fusca. The episclera is composed of loose bundles of collagen and elastic fibers, blood vessels, nerves, fibroblasts and, depending on the species and color pattern, melanocytes. Like the corneal stroma that of the sclera consists of collagen lamellae but unlike the cornea they are obliquely arranged and lack uniformity. This lack of uniformity, along with the variability in collagen fiber diameter and the relative deficiency in water-binding glycosaminoglycans's, account for the lack of transparency to the sclera. The sclera possesses several channels or emissaria to allow the passage of nerves and blood vessels. Examples of these are the optic nerve, the vortex veins, long posterior ciliary arteries, and the long and short ciliary nerves. These emissaria are weak points and are potential sites of rupture following a trauma.

Vascular Tunic

The vascular tunic is comprised of three distinct regions, (1) the iris, (2) the ciliary body, and (3) the choroid. The vascular tunic is mesodermal in origin and is situated between the outer fibrous tunic and the inner nervous tunic. The vascular tunic is also refered to as the uvea.

Iris

The iris is the anterior most portion of the vascular tunic and functions as a moveable diaphragm between the anterior and posterior chambers. Embryologically, the iris is comprised of components that originate from neural ectoderm, the sphincter and dilator muscles and the 2-layered posterior epithelium, and mesodermal components that form the iris stroma. The pupil varies in shape between species and can be circular (dog, rabbit), oval (horse, cow), or vertical (cat). This central portion of the iris normally rest against the anterior lens surface and in the absence of this support, the iris will tremble. The color of the iris can vary with species, age, and sex, and can even differ between eyes or within eyes in the same animal. The color of the iris depends on the pigmentation of both the stromal melanocytes and the posterior epithelium. In animals with blue eyes the pigment is absent in the stroma, while in the albino it is lacking in both the stroma and the posterior epithelium.

The anterior border of the iris is often discussed as possessing an epi- or endothelial cell layer. This is incorrect. The anterior border is formed by fibroblasts and melanocytes. The peripheral portion of the iris inserts as the pectinate ligaments in the area where Descemet's membrane ends. This angle formed by the cornea and the root of the iris is termed the iridocorneal angle and is the entrance to the trabecular meshwork, the site of outflow of the aqueous humor.

The iris stroma is a loose tissue comprised of collagen bundles, blood vessels, nerves, melanocytes, fibroblasts, and the sphincter muscle of the iris. The blood vessels are endothelial lined structures with a basement membrane, smooth muscle, and pericytes. The capillary endothelium is not fenestrated and is part of the blood-eye-barrier.

The sphincter muscle is found within the iris stroma. It is smooth muscle in mammals and striated in birds and lower vertebrates. As stated previously, it has its origin from neural ectoderm. It is situated in the pupillary zone and varies in shape depending on the shape of the pupil. Contraction of the sphincter muscle results in a decrease in the pupil diameter, termed miosis. The sphincter muscle is innervated by parasympathetic fibers from the Edinger-Westphal nucleus, the fibers of which travel with the third cranial nerve (oculomotor).

The posterior epithelium of the iris is actually 2 cell layers, both of neuroectodermal origin. Because of their origin from the invaginating optic cup these cells are situated apex to apex and remnants of the optic vesicle may occasionally persist between these cell layers. The posterior most cell layer is pigmented and is continuous with the non-pigmented layer of the ciliary body and ultimately with the neural retina. It is separated from the posterior chamber by a basal lamina. Adjacent, posterior epithelial cells are attached by desmosomes and terminal bars to each other. The anterior epithelial layer consists of a pigmented apical portion and a myoepithelial basal portion which forms the dilator muscle of the pupil. The dilator muscle is a radially-arranged, smooth muscle and is innervated primarily by sympathetic fibers.

In herbivores a structure termed the corpora nigra or granula iridica is found. This an extension of the posterior epithelium into the pupil both dorsally and ventrally. These appear as 3-4 pigmented, irregular masses. The superior ones are usually larger. They are a normal finding.

Ciliary Body

The ciliary body, like the iris, contains both neurectodermal and mesodermal tissue. It is divided into 2 parts, the anterior pars plicata and the posterior pars plana. In saggital section the ciliary body is triangular in shape with its base at the iris and the apex posteriorly at the ora ciliaris retinae. The pars plicata possesses 70-100 major ciliary processes and between these ciliary processes are valleys in which can be found smaller minor processes. Each ciliary process contains a vascular core and is covered by a 2-layer epithelium. The ciliary processes are the source of production of the aqueous humor.

The posterior two layered epithelium of the iris continues over the ciliary body as the outer pigmented and inner non-pigmented epithelia. As is the case in the iris, these epithelial cells are arranged apex to apex. At the junction of the ciliary body and the retina, the single layered non- pigmented epithelium (NPE) continues as the multi-layered retina and the pigmented epithelium of the ciliary body continues as the retinal pigment epithelium. Unlike in the retina and the iris, the two cell layers of the ciliary epithelium are strongly attached to each other by terminal bars. In addition, there are zonulae adherentes and zonulae occludens between the apices of the non-pigmented cells. This is the site of the ciliary body blood-aqueous-barrier.

The lenticular zonules that support the lens originate in the pars plana and travel forward in the valleys between the ciliary processes. The zonules ensheath the ciliary processes as they insert on the lens capsule both anterior and posterior to the lens equator. The zonules appear to be elastic microfibrils.

The pigmented cells are joined to each other by desmosomes and the intercellular space of these cells is therefore permeable. The basal portion of the cell faces the ciliary body stroma and is covered by a basal lamina. Beneath the pigmented cell layer is the stroma of the ciliary body. This is mesodermal in origin and contains vessels, nerves, collagen bundles, smooth muscle, melanocytes, and fibroblasts. The stroma of the ciliary processes is a highly vascularized connective tissue with capillaries containing large fenestrations, 300-1000A. The ciliary muscle is smooth and in all primates is comprised of 3 parts: the longitudinal, circular, and radial portions. This muscle is poorly developed in most other mammals, with the longitudinal portion predominating. This muscle is primarily innervated by parasympathetic fibers, but also has some adrenergic innervation. This is the muscle of accommodation and serves to change the shape of the lens.

Choroid

The choroid is the third component of the uvea. It extends from the edge of the optic nerve to the pars plana. The choroid is loosely attached to sclera. The choroidal stroma varies from being darkly pigmented to a complete absence of pigmentation depending on the species, breed, and individual. In those animals lacking pigmentation, the fundic reflex appears red and on ophthalmoscopic examination the choroidal vessels are visible. The stroma itself is composed of collagen fibrils, melanocytes, fibroblasts, nerves, and larger blood vessels. The majority of the vessels are veins with arteries situated amongst them.

The capillary layer of the choroid, the choriocapillaris, is found in the inner portion just below the retinal pigment epithelium (RPE). It is responsible for the nutrition of the RPE and the outer retina and, in some species, for the entire retinal nutrition. The capillaries form a lobular network and are comprised of typical fenestrated endothelial cells surrounded by a basal lamina that surrounds and is shared by pericytes and smooth muscle cells. These lobules are supplied by a central arteriole and are surrounded by a ring of postcapillary venules. Venous drainage from the choroid occurs in four quadrants where blood collects in an ampulla and then drains via one of four vortex veins that penetrate the sclera.

In addition to above structures, some species also contain a layer in the choroid, the tapetum lucidum, situated between the larger choroidal vessels and the choriocapillaris. Because of its location, the vascular communications between the larger vessels and the choriocapillaris must traverse the tapetum. The tapetum is responsible for the so called "eye-shine" or bright, colored reflection seen in certain species. It is situated in the superior one-half to one-third of the choroid. The color of the tapetum varies with and within species and can be green, blue, yellow, orange, or a variation of these. The tapetum is found in dogs, cats, ferrets, horses, ruminants, and nocturnal animals. It is absent in humans, other primates, pigs, rabbits, rats, guinea pigs, and mice. The tapetum can be cellular (dog, cat, ferret, and other carnivores) or fibrous (horse, ruminants, and other ungulates).

Nervous Tunic

The retina is the most complex of all the ocular tissues. The classic description of the retina includes 10 layers that, from the outside in, are as follows:

1. Retinal pigment epithelium (RPE)
2. Photoreceptor outer segments – Rods and Cones
3. Outer limiting membrane
4. Outer nuclear layer
5. Outer plexiform layer
6. Inner nuclear layer
7. Inner plexiform layer

8. Ganglion cell layer
9. Nerve fiber layer
10. Inner limiting membrane

Embryologically, the retina is neuroectodermal in origin and forms first as an optic vesicle and later invaginates to form the optic cup. The posterior wall of the optic cup gives rise to the retinal pigment epithelium (RPE), while the anterior wall gives rise to the remaining 9 layers of the retina. The space between these layers that was present embryologically remains as a potential space and it is here, between the RPE and the sensory retina, that retinal detachments usually occur. In humans, the retina is mature at birth but, in many other species, the development of the retina continues postnatally. In the dog and cat, the retina is not mature until approximately 3-5 weeks centrally and 8-9 weeks peripherally.

Retinal pigmented epithelium (RPE)

The RPE is the homolog of the epithelium of the choroid plexus of the brain. It is a monolayer of hexagonal cells which is continuous anteriorly with the pigmented epithelial layer of the ciliary body and the anterior epithelium of the iris. External to the RPE is Bruch's membrane and the choriocapillaris. The inner apical surface of the RPE has numerous long villous processes that surround the outer segment of the photoreceptors. There is no attachment between these cells, but an acid mucopolysaccharide ground substance exists between them. The lateral RPE cell margins, near the apical surface, have terminal bar attachments (zonula occludens, zonula adherens) which are one portion of the blood-retinal barrier. As the name suggests, the RPE contains melanin granules. These are found predominantly in the inner third of the cytoplasm. An important exception to this is seen in animals possessing a tapetum where the RPE overlying the tapetal portion of the choroid is non-pigmented.

Sensory Retina

The sensory retina is thickest near the optic nerve and thins towards the periphery. External to the sensory retina is the RPE and internally is the vitreous humor. The photoreceptors are in the outer portion of the sensory retina and can be divided into the cell body, inner segment, and outer segment. In addition, photoreceptors are divided into rods and cones based on their morphology, physiology, and sensitivity. As suggested by their name, the rods are long and slender, while the cones tend to have a shorter, wider appearance. The cones are further divided according to their spectral absorption characteristics into the red-sensitive (570 nm), green-sensitive (540 nm), and the blue-sensitive (440 nm) cones.

Photoreceptors (first neuron layer)

This is the outermost cells of the sensory retina and is also known as the rod and cone layer.

In this layer, a photochemical process takes place to convert light to an electrical impulse occurs in the outer segment. The nuclei of the photoreceptor cell forms the outer nuclear layer and the axons of the photoreceptors synapse with processes of the bipolar neurons in the outer plexiform layer

The rods are highly sensitive to light. They provide minimal detail and are inactive during daylight vision. The rods predominate in all portions of retina (except the fovea of birds and primates)

The cones are less sensitive to light. They provide more detailed vision, color vision and predominant only in the fovea.

Integrating neurons (second neuron layer)

The integrating neurons receive input from the photoreceptors and transmit it to the ganglion cells. The nuclei of these cells form the inner nuclear layer.

Ganglion cells (third neuron layer)

The ganglion cells receive input from the integrating neurons and transmit impulses to the brain. The axons of the ganglion cells form the nerve fiber layer and converge at the posterior pole of the eye to form the optic nerve.

Muller cells

These are the supporting cells of the retina, however they do not participate directly in the electrical impulse transmission. The nuclei of Muller cells lie in the inner nuclear layer among the nuclei of the bipolar neurons.

Retinal blood vessels

The microanatomy of retinal blood vessels varies according to species. Some species lack retinal blood vessels entirely. The retinal blood vessels are located in the nerve fiber layer and nourish the only inner third of the retina. The outer 2/3 of retina nourished by choroidal vasculature. The pattern of retinal vessels varies among species, but in all species retinal vessels originate from or directly adjacent to the optic nerve.

Vision

Vision is a complex process that involves a photochemical reaction converting light into electrical impulses. Light photons enter the eye through the cornea and pass through the transparent media of the aqueous humor, the lens and the vitreous body. After passing through the sensory retina and the retinal photoreceptors. Some of these photons are absorbed and converted to an electrical potential. The remaining light passes through the RPE and choriocapillaris to the tapetum. The tapetum acts to reflect light back to the retina and allow more photons to be absorbed by photoreceptors and converted to an electrical potential.

These electrical impulses are transmitted to the bipolar neuron, modified then transmitted to the ganglion cell. The axons of ganglion cells become part of the optic nerve. The optic nerve perforates the sclera at the lamina cribrosa and proceeds toward the brain. The two optic nerves meet at the optic chiasm.

Certain species possess a highly specialized area of the retina, the macula, which contains the fovea. This is found in humans, primates, birds, and some fish and reptiles. In addition, an area centralis is described in several other laboratory animals such as the cat, but these animals do not possess a true fovea. In the center of the macula is an area, the fovea, which is the area of highest visual acuity. It is completely rod free, containing only cone photoreceptors. The fovea is an avascular area and lacks all retinal layers except the photoreceptors and their nuclei.

The Lens

The lens is a biconvex transparent structure and is the second refracting unit of the eye. It lies posterior to the iris and is suspended from the ciliary body by the zonular fibers. It also has a posterior attachment to the anterior vitreous face where it lies in a depression of the vitreous, the patellar fossa. It is a unique tissue in that it is avascular, transparent, lacks nerve supply, and has the highest concentration of protein in the body. Embryologically, the lens originates from the surface ectoderm that is induced to form the lens placode and invaginate by the advancing optic vesicle and cup.

The lens is surrounded by a basement membrane, the lens capsule, which is secreted by the lens epithelial cells anteriorly and the cortical fibers posteriorly. The capsule is thickest anteriorly and thinnest at the posterior pole and continues to grow throughout life.

Beneath the lens capsule, anteriorly, is the lens epithelium. These form a monolayer of cuboidal cells whose apices face the cortical fibers with the basal portion of the cell adjacent to the lens capsule. The apical portion of the epithelial cells has terminal bars. Adjacent to the lens equator is the pre-equatorial or germinative zone where the lens epithelial cells replicate. It is these cells that are susceptible to radiation and toxic insult, resulting in cataract formation. The newly formed cells migrate equatorially where they elongate, differentiate into cortical fibers, are displaced inward, compressed, and loose their nuclei. This replicative process occurs throughout life. As these cortical fibers elongate anteriorly and posteriorly they attach at a line, the lens suture. In human, dogs, and cats these suture lines are in a Y-shape, upright anteriorly and inverted posteriorly, but this varies with other species. This results in a layered arrangement of the lens fibers with the oldest fibers in the center and the newest fibers surrounding them. On cross-section, the lens fibers are hexagonal and adjacent cells interdigitate with each other in the form of microplicae and by a ball-and-socket arrangement.

The Vitreous

The vitreous is a semifluid and transparent structure composed of 99% water with the remnants composed of collagen fiber matrix and mucopolysaccharide (hyaluronic acid). It is normally avascular and does not contain nerves or lymphatics. The vitreous is attached anteriorly to the zonules and to the posterior lens and peripherally at the pars plana and around the optic nerve.

Primary vitreous (the hyaloid vessel system) forms as the lens vesicle and optic cup ar forming. The hyaloid artery originates from the site of the future optic nerve and extends to the posterior aspect of the developing lens. This system is compressed centrally and gradually atrophies as the secondary vitreous forms. It has usually disappeared completely by 2-3 weeks of age in the dog but portions may persist for a few months in ruminants.

Secondary vitreous is produced by the Muller cells of the retina. As it forms, and the eye enlarges, the primary vitreous becomes proportionally smaller and axially located. The secondary vitreous becomes the adult vitreous.

Tertiary vitreous consists of the ciliary zonules (lenticular zonules) suspending the lens. It forms from the neural ectoderm (surface epithelium) of the ciliary body and condenses into the individual zonules.

Optic Nerve

The optic nerve is formed by axons of the nerve fiber layer. These nerve fibers develop from the ganglion cells to form the nerve fiber layer of the retina, extend posteriorly to form the optic nerve and extent toward the brain. Myelin develops from the brain outward to reach the eye a few weeks after birth. In the dog, myelin usually extends into the eye and a short distance along the nerve fiber layer. In the cat, myelin usually does not enter the eye. Dura mater and arachnoid of the brain extend along the optic nerve to blend into the posterior sclera.

Appendix 1: General Principles Of Histotechnology

The simplest method for the direct observation of living cells is to remove them from the organism, place them on a slide and examine them by dark-field or phase-contrast microscopy or light microscopy after staining.

However, living tissues and cells are difficult to examine microscopically; instead, they are usually killed by careful fixation to minimize alterations of in vivo morphology. Subsequent processing ends with tissues embedded in a material which facilitates thin sectioning; for light microscopy this is usually paraffin. The steps for processing tissue for histology are as follows.

- **Removal** of the tissue.
- **Fixation**. The sample, preferably small, is placed in a reagent (e.g. formalin which is 10% aqueous solution of 40% formaldehyde) which preserves substances and structures in the cells and tissues and prevents autolytic changes. Depending on the size of the piece, it should be fixed from 4-24 hours.
- **Washing and Dehydration**. After fixation the specimen is washed in water to remove the excess fixative then placed in increasing strengths of alcohol or other dehydrating agents. (So the original water in the tissue and what was in the water and not precipitated by the formalin can be removed by the dehydrating agent – eg. edema fluid with low level albumin/ protein).
- **Clearing.** This is the process of removing the dehydrating agent with some fluid which is miscible both with the dehydrating agents and with the embedding medium. (Since a common embedding agent is paraffin which is an oil, clearing agents often replace oils that are not precipitated by formalin – eg. triglycerides in fat cells).
- **Infiltration and Embedding**. In this process the clearing agent is replaced with the embedding medium, usually paraffin.
- **Cutting.** Slicing and removing of 3 and 10 micrometers (microns) – thick sections of the tissues with a microtome.
- **Section Mounting**. It is accomplished by transferring the sections to a clean glass microscope slide.
- **Deceration.** This is the removal of the embedding agent.

❖ **Staining.** This is a process of increasing the visibility of cells by the application of dyes or by their reaction with chemical agents to form visible substances.

Mounting. This is the act of covering the cleared section with a drop of mounting medium and a thin glass cover-slip. Mounting media have the same indices of refraction as glass and harden as the solvent evaporates, thus making the preparation permanent.

Appendix 2: Histological Staining Methods

Staining brings out in contrasting colors otherwise colorless structural details of tissue sections. The coloring ability of many dyes resides in either their acid or basic radical. If in the anion, the dye is called **acid**, if in the cation, it is called **basic**.

Acid components of a cell (like nuclear chromatin or chromosomes) stain with basic dyes (e.g., hematoxylin) and these components are referred to as **basophilic** or **hematoxylinophilic** (blue).

Basic components (various kinds of cytoplasm and intercellular substance) take acid dyes (e.g., eosin), and are called **acidophilic** or **eosinophilic** (pink). Please note that there are numerous stains used to evaluate tissue for different purposes. Hematoxylin and eosin (or H&E) staining is the most common method of staining used for histologic sections.

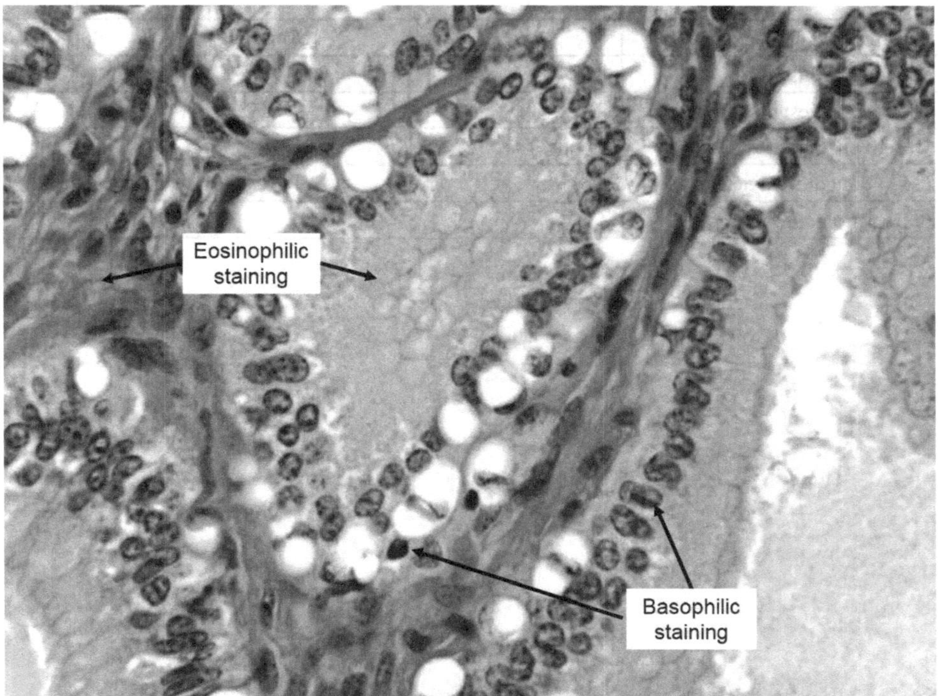

Basophilic and eosinophilic staining in a tissue section.

Appendix 3: Measurement For Microscopy

Though the millimeter is a convenient unit for measurement with magnification, it is too large for measuring microscopic objects conveniently. For microscopic measurement the millimeter is divided into 1000 equal parts, each part being called a micrometer or a micron. The symbol is um. A micrometer, then, is one millionth part of a meter. The smallest object under the most favorable background conditions that can be observed with the unaided eye is of the dimensions of 0.1 millimeter or 100 um.

Under the ideal background conditions the human ovum and the largest of nerve cells are barely visible. Objects which are below the size for visual examination with the unaided eye and are yet large enough to be seen with a bright field microscope are said to be microscopic. The range is from 0.1 millimeter to 0.0001 mm or from 100 um and 0.1 um. Particles from 0.1 um to 0.004 um in size are called submicroscopic (or ultrastructural) particles.

A still finer unit of measurement is the **angstrom unit** (A). Its value is 0.000,000,1 of a mm. The angstrom unit has recently been abandoned in favor of the nanometer, but many references to it remain.

	Millimeters (mm)	**Micrometers (μm)**	**Nanometers (nm)**
Ovum	0.1	100	100000
Red Blood Cell (canine)		7.0	7000
Staphylococcus		0.8	800
Viruses (largest pox)		0.1	100
Casein particles in milk		0.15	150
Colloidal gold particles			15
Starch molecule			5

Index